Practical Applications of Asymptotic Techniques in Electromagnetics

DISCLAIMER OF WARRANTY

For a listing of recent titles in the *Artech House Electromagnetics Series*,
turn to the back of this book.

Practical Applications of Asymptotic Techniques in Electromagnetics

Francisco Saez de Adana
Oscar Gutiérrez
Iván González
Manuel F. Cátedra
Lorena Lozano

ARTECH HOUSE
BOSTON | LONDON
artechhouse.com

Library of Congress Cataloging-in-Publication Data
A catalog record for this book is available from the U.S. Library of Congress.

British Library Cataloguing in Publication Data
A catalogue record for this book is available from the British Library.

ISBN-13 978-1-60807-063-3

Cover design by Vicki Kane

685 Canton Street
Norwood, MA 02062

10 9 8 7 6 5 4 3 2 1

Contents

Preface

This book explores the applications of asymptotic numerical techniques to the analysis of real-world engineering problems. The content of this book is based on our experience as users and especially as developers of computer tools for analyzing problems of radiation, propagation, and scattering in the high-frequency range or in the range where the size of the object under analysis is larger than the wavelength. After several years of working with these electromagnetic techniques and developing multiple computer programs based on them (some of them included in the companion DVD for this book), we decided to express these experiences in a book that would be beneficial for readers who are interested in implementing these techniques to analyze engineering problems. With this aim, we have given a practical focus to this book to demonstrate the capabilities of asymptotic methods in the electromagnetic analysis of realistic problems.

Historically, asymptotic techniques have been widely used for high-frequency problems. These techniques are applied to problems such as the analysis of on-board antennas, the computation of the radar cross section (RCS), antenna design, or propagation in mobile communications. However, the accuracy of these techniques is related to the fidelity of the geometric modeling compared to the real scenario. There are two options when the engineer starts to deal with this kind of problem. The first option is a simplification of the modeling using canonical structures to simplify the geometric treatment associated with the asymptotic techniques. The disadvantage of this option is a loss of accuracy in the results. The second option is an exact modeling of the scenario to obtain results that are as accurate as possible. In this case, the price is the computational cost associated with the solution. Our objective with this book is to show how this second modeling option can be applied to obtain very accurate results; combining electromagnetic methods with some geometrical techniques enables the solution of realistic and complex problems with a reasonable amount of computational resources.

With this objective, we cover the most widely used asymptotic techniques, such as geometrical optics and the uniform theory of diffraction (GO/UTD) and physical optics and the physical theory of diffraction (PO/PTD), whose foundations are thoroughly described in Chapters 3 and 4. We also show how their electromagnetic formulation must be combined with the proper geometrical modeling to tackle real-world engineering problems. For this reason, Chapter 2, which describes the modeling of complex objects by parametric surfaces called nonuniform rational B-splines (NURBS), is included because, based on our experience, this technique is one of the most suitable for combining electromagnetic analysis with an accurate description of the real object. Therefore, in Chapters 3 and 4, in addition to the electromagnetic formulation itself, its application to objects modeled by NURBS is also included. However, the proposed analysis would not be possible in a real engineering environment if they were not computationally affordable. For this reason, Chapter 5 is focused on the application of acceleration techniques to reduce the computational costs associated with the analysis of complex objects modeled by NURBS.

These theoretical concepts would remain purely academic, however, if examples of real problems were not included in this book. The capabilities of the presented techniques are described in Chapter 6 through examples of real engineering problems that we have analyzed over the years, showing how complex problems can be treated using a proper implementation of asymptotic techniques with a low computational cost. As a benefit to the reader, a demo version of the codes with which these cases have been analyzed is included in the companion DVD of this book to allow engineers to build and analyze their own cases with the help of the user's guide also included in the DVD. The only difference between the DVD version and the commercial codes is the number of surfaces that can be analyzed. In this case this number is limited, although it is still enough to allow the reader to perform sufficiently complex analysis to become familiar with the techniques presented in the book.

Therefore, this book is targeted towards engineers and researchers working on antenna analysis and design in the fields of radiation, propagation, and scattering. This book can also be used by graduate students in the fields of electromagnetism, antennas, propagation, or radio communication systems, although basic knowledge of electromagnetic theory, antennas, and propagation is assumed.

We would like to acknowledge the invaluable help provided by many people in the development and execution of this book. First of all, we thank all the members of our team who have contributed, in the past or in the present, to the development of the techniques presented in the book: Mara Jesús Algar, Olga María Conde, Carlos Delgado, Eliseo García, Josefa Gómez, José Manuel Gómez, Jesús Pérez, Fernando Rivas, and Abdelhamid Tayebi. We are also grateful to all the companies and institutions that have, throughout the years, trusted our programs for their analyses, especially of the cases included in Chapter 6: the Centre National d'Études Spatiales (CNES, French) for the Jason and Stentor cases, the European space agency (ESA) for the International Space Station (ISS) case, GMV (Spain), ESA and the Institute of applied systems technology, JOANNEUM RESEARCH (Graz, Austria) for the Vienna cases, ACCIONA for the wind generator case and the Spanish defense department for the Tabarca case. We also want to thank Artech House for its confidence in our work and its reviewers and editors for their helpful commentaries during the development of this book.

Finally, we do not want to forget to be sincerely grateful to our families for all the support during these years, especially in those moments when our dedication to asymptotic techniques have stolen some of the time that could have been dedicated to them. This book would not be possible without them.

Francisco Saez de Adana
Oscar Gutiérrez
Iván González
Manuel Felipe Cátedra
Lorena Lozano
Alcalá de Henares, Spain
October 2010

Chapter 1

Introduction

1.1 ASYMPTOTIC TECHNIQUES AND REAL-WORLD PROBLEMS

The numerical techniques for electromagnetic analysis have been in great demand for engineering applications for several decades. Such techniques are essential for many problems and include analyzing on-board antennas, studying propagation to characterize the radio channel for mobile communication applications, or predicting the radar cross section (RCS) of complex targets.

In the first two problems, the behavior of an antenna is affected by the objects surrounding it. Because antenna efficiency is critical when the antenna is installed in a complex system, prior knowledge of the level of degradation of its free space radiation pattern is very important. Thus, there are several methods that provide the necessary information regarding the behavior of the antenna once it is installed in the environment where it will operate to the system designer. In general, the same can be stated about the necessity of predicting the interaction between an electromagnetic field and a complex structure when designing systems in which the response to an electromagnetic field can be an important feature, such as in the design of aircrafts, tanks, or ships with respect to their responses to radar systems, especially for military applications.

When the interaction between an electromagnetic source and a complex structure is analyzed, there are two alternatives: perform measurements of the structure (using the real structure or a scaled model) or calculate the interaction between the antenna and the structure using a computational model. The measurements, although they can provide very good predictions if they are correctly performed, have several disadvantages that are mostly related to the high economical cost associated with the measurement process. Both working with the real structure and building an accurate mock-up that represents it have a high cost, which must be added to the cost associated with the maintenance of the facility where the measurements are performed.

On the other hand, computing the interactions within an electromagnetic field structure presents an important number of advantages: this interaction can be performed prior to designing the system, which allows some problems to be identified in advance, and in the case of analyzing antennas mounted on structures, it allows different antenna locations to be investigated. If the numerical tool is efficient enough, the cost in terms of time and money is lower than that from any kind of measurement. Obtaining this computational efficiency has promoted the development of different electromagnetic techniques that are applied to different problems depending on factors that include the structure size or the frequency of analysis.

The different electromagnetic techniques are based on Maxwell's equations, and they can be classified following different criteria. One very popular classification is to divide the electromagnetic techniques into two large families: rigorous techniques and asymptotic techniques. The rigorous techniques, such as the method of moments (MoM) [1], the finite difference time domain (FDTD) [2], or the finite element method (FEM) [3], derive an exact numerical solution of the problem. However, these techniques are based on the discretization of the geometry in differential elements with sizes of approximately $\lambda/8$ or $\lambda/10$. Therefore, for larger structures relative to the size of the wavelength, the computational resource requirements increase, and these techniques become limited by the availability of computational resources. In the past decade, a strong emphasis has been placed on developing rigorous techniques with new formulations based on multipoles and multilevels, such as the fast multipole method (FMM) and the multilevel fast multipole method (MLFMM) [4]. These developments, together with the intensive use of the parallel computation using large computer clusters, have allowed rigorous techniques to be applied to solve problems in the microwave range of frequencies. Obviously, this has significantly advanced the application of rigorous techniques to electrically large problems, but they are still not applicable to the design of large arrays or platforms because their design assumes an optimization iterative process, which requires a high amount of analysis that includes high computational cost.

On the other hand, the asymptotic techniques approximate Maxwell's equations to a solution whose accuracy increases with frequency. The main advantage of these methods is that the computational cost remains constant as the frequency increases, while, as has been mentioned, the accuracy of the result increases. Therefore, these methods are especially suitable for analyzing electrically huge problems, particularly in the design process where the number of analysis can be really high, as mentioned earlier. The most popular techniques among the asymptotic techniques are geometrical optics and the uniform theory of diffraction (GO/UTD) [5, 6] and physical optics and the physical theory of diffraction (PO/PTD) [7, 8]. This book is focused on these kinds of techniques.

However, when the problem of applying these electromagnetic techniques to the analysis of realistic problems is considered, the geometric modeling of the structure to be analyzed is as important as the implementation of an efficient numerical technique. Not only does geometric modeling have to accurately describe the real structure to improve the quality of the predictions, but the information associated with the model must also be stored efficiently. If a complex body can be represented using a small amount of information, its interaction with the source is more efficiently computed and the computational cost associated with the analysis is considerably reduced. Therefore, two main factors must be considered regarding the geometric modeling: the accuracy of the representation and the efficiency of the data storage that is necessary to achieve this accuracy.

Initially, when the application of electromagnetic numerical techniques to complex problems was first considered, geometric modeling was accomplished by using simple shapes to model complex bodies. A complex platform was modeled using canonical structures, such as cylinders, spheres, and plane facets. However, as the problem to be analyzed became increasingly complex, two difficulties arose: the representation of the real structure using canonical shapes was not accurate enough, and the number of canonical shapes necessary to represent a real body increased. Thus, parametric surfaces were introduced and have become very popular recently. The use of parametric surfaces has several advantages. First, they can represent a complex body with a high level of accuracy using a small amount of information, which provides very good predictions in the application of electromagnetic techniques to this kind of modeling. Moreover, the application of the parametric representation to the electromagnetic analysis gives the designer the perfect link from the electromagnetic world to other worlds, where the parametric representation is widely extended to aeronautic and even the architectural applications. Therefore, it is obvious that this kind of modeling naturally allows integration of the electromagnetic analysis in the design process because there is a link between the electrical engineer with other members involved in the design team, such as an aeronautical engineer. Because the main objective of this book is to provide the electrical engineer with a guide to the application of asymptotic electromagnetic techniques for the analysis of real-world problems, a parametric representation of the body is an essential part of the book.

1.2 STRUCTURE OF THIS BOOK

Chapter 2 presents the geometric representation of a complex body that is necessary for further application of numerical electromagnetic techniques to real-world problems. The chapter is focused on the parametric representation of complex bodies as explained in Section 1.1. Specifically, a representation using nonuniform rational B-splines (NURBS) is described because this type of surface has become one of the most suitable for electromagnetic analysis after it gained popularity in the world of computer-aided graphic design (CAGD) and its extensive application in the world of the aeronautical design. A description of NURBS surfaces is provided in Chapter 2, as well as a procedure to compute the necessary parameters to apply asymptotic techniques over NURBS surfaces.

Chapters 3 and 4 address asymptotic techniques and present GO/UTD and PO/PTD, respectively. Other books cover these formulations in detail. Clearly the classical formulation must be presented in this type of book, and therefore both chapters present a review of those formulations. However, taking into account the main aim of this book, which is to specifically address the application of both techniques to realistic bodies modeled by parametric surfaces, only an overview is presented. Therefore, aspects of ray-tracing over parametric surfaces are treated in Chapter 3, and the computation of the PO integral over this type of surface is developed in Chapter 4. Moreover, both chapters include an historic overview of the application of asymptotic techniques to real-world problems.

All these tasks that must be performed when analyzing complex bodies outside the electromagnetic computation itself are the most expensive from a computational point of view. Therefore, efficient tools based on asymptotic techniques, the acceleration techniques, must be included in the model. The acceleration techniques are based on the approaches used in computer graphics to solve visualization problems, that is to say, to determine which parts of a scene are visible or not visible from a given source (typically the eye of the observer). The application of these acceleration techniques to the electromagnetic analysis world is presented in Chapter 5. There are several applications of such techniques to plane facets models in the literature, but this book focuses more intensely on the application to parametric surface modeling.

Finally, Chapter 6 presents some results of the application of asymptotic techniques to real-world problems. Radiation, propagation, and scattering problems are depicted. The validity of the asymptotic techniques by comparison with measurements or rigorous techniques is shown. Moreover, a discussion of the main features of the application of such techniques to this kind of problem is included in the chapter.

References

[1] R. F. Harrington. *Field Computation by Moment Methods.* MacMillan, 1968.

[2] A. Taflove. "Application of the finite-difference time domain method to sinusoidal steady-state electromagnetic-penetration problems." *IEEE Transactions on Electromagnetic. Compatibility,* 22(3):191–202, August 1980.

[3] O. L. Zienkiewichhz. *The Finite Element Method.* McGraw-Hill, 1979.

[4] W. C. Chew, J. Jin, E. Michielssen, and J. Song. *Fast and Efficient Algorithms in Computational Electromagnetics.* Artech House, 2001.

[5] R. G. Kouyoumjian and P. H. Pathak. "A uniform geometrical theory of diffraction for and edge in a perfectly conducting surface." *Proceedings of the IEEE,* 62(11):1448–1461, November 1974.

[6] D. A. McNamara, C. W. I. Pistorius, and A. G. Maherbe. *Introduction to the Uniform Geometrical Theory of Diffraction.* Artech House, 1990.

[7] A. Straton. *Electromagnetic Theory.* McGraw-Hill, 1941.

[8] P. Ya. Ufimtsev. "Elementary edge waves and the physical theory of diffraction." *Electromagnetics,* 11(2):125–160, April-June 1991.

Chapter 2

Geometric Description of the Bodies and Computation of Geometric Parameters

2.1 OVERVIEW OF THE GEOMETRIC MODELING OF COMPLEX OBJECTS

The application of numerical techniques to solve realistic problems in electromagnetism requires proper geometric modeling. The efficiency of the geometric treatment is as important as the efficiency of the electromagnetic methods implemented. Thus, geometric modeling of complex bodies for applications in computational electromagnetics has evolved to increasingly exact and accurate representations of reality.

Six big groups of methods for modeling solid bodies have been used: parametric shapes [1], cell decomposition [2, 3], constructive solid geometry (CSG) [3, 4], boundary representation [1, 2, 5], and wireframes [5]. Each modeling method presents advantages and disadvantages compared with the others. Therefore, choosing one of them depends on the application in which it will be used and the necessary requirements.

In the previously mentioned electromagnetic problems, the boundary representations have been the most widely used and, more specifically, the boundary representation with plane facets. This method is based on an approximation of the object's surface with a set of plane facets and straight edges that connect the different facets of the model [6, 7]. This modeling technique has been used successfully to solve electromagnetic problems using both rigorous techniques and asymptotic techniques, mostly because of its simplicity. However, it presents an important number of disadvantages because of the high number of facets required to model a complex body, the error committed by approximating a curved surface by a set of plane facets, and the appearance of the false edges that must be taken into account. Figure 2.1 shows an Airbus A320 aircraft modeled by 5,625 facets.

Figure 2.1 Airbus A320 aircraft model using 5,625 facets.

It is obvious that, for a complex body as that shown in Figure 2.1, geometric modeling using curved surfaces is more suitable; therefore, the use of a boundary representation using curved parametric surfaces has been extended [8–10]. For instance, in the application of asymptotic techniques to electromagnetic problems, such as analyzing on-board antennas and computing the RCS of complex bodies, the representation using NURBS surfaces instead of plane facets is growing very quickly [11–15].

The main advantage of this technique is that it is possible to represent any kind of surface with a small amount of information compared to other techniques. Currently, most of the modern computer-aided graphic design (CAGD) packages, such as Microstation [16] or Rhinoceros [17], describe bodies in terms of NURBS because they are widely used in other fields, including aeronautics. One proof of its popularity is the fact that NURBS surfaces are standard entities in the initial graphics exchange specification (IGES) format [18], which is one of the most common formats for representing realistic bodies geometrically. Figure 2.2 shows the description of a NURBS surface in IGES format.

The description using NURBS surfaces provide an accurate representation of a real model with less information than that necessary for a plane facets model. Moreover, all the problems of the plane facets representation are avoided. Figure

```
128,1,1,1,1,0,0,1,0,0,0.0D0,0.0D0,0.11D0,0.11D0,0.0D0,0.0D0,          0000025P   13
0.11D0,0.11D0,1.0D0,1.0D0,1.0D0,1.0D0,0.0D0,0.055D0,-0.055D0,         0000025P   14
0.0D0,-0.055D0,-0.055D0,0.0D0,0.055D0,0.055D0,0.0D0,-0.055D0,         0000025P   15
0.055D0,0.0D0,0.11D0,0.0D0,0.11D0;                                    0000025P   16
```

Figure 2.2 Description of a NURBS surface in IGES format. Entity 128.

2.3 shows the same model as in Figure 2.1, but in this case, only 620 NURBS surfaces are required. However, the increased accuracy requires higher computational time for the geometric treatment that the asymptotic techniques require. This disadvantage can be overcome using acceleration techniques that will be presented in Chapter 5.

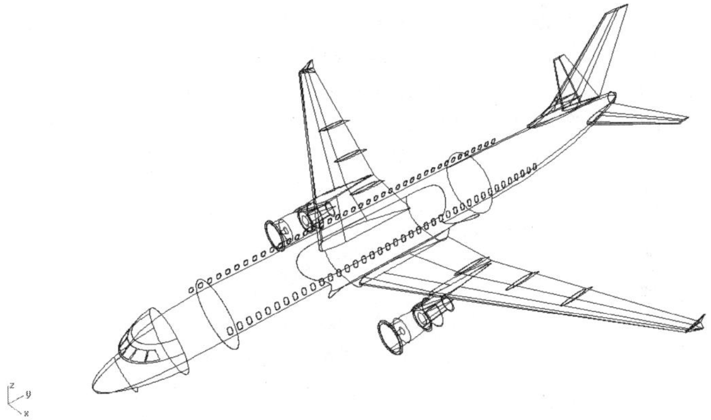

Figure 2.3 Airbus A320 aircraft model using 620 NURBS surfaces.

As it has been noted, the NURBS format is widely used because of its accurate representation of the geometry with a small amount of information. However, when applying asymptotic techniques to a NURBS representation, some geometric parameters must be computed. NURBS surfaces cannot be used directly for this computation because of its numerical instability. Thus, when applying the NURBS representation in the complex problems analysis, NURBS surfaces are decomposed into Bezier patches, which are numerically more stable. This decomposition is performed using the Cox de Boor algorithm [19]. Therefore, internally, the geometric treatment is carried out using Bezier surfaces.

This chapter presents the description of both Bezier and NURBS surfaces. With this aim, the basis on Bezier and NURBS curves is presented because the description of both kinds of surfaces is derived by the curves representation. Moreover, this chapter describes the Cox de Boor algorithm used to decompose NURBS surfaces in Bezier surfaces. Finally, the procedure to compute some geometric parameters over Bezier surfaces is provided at the end of this chapter.

2.2 RATIONAL BEZIER CURVES

The Bezier curves were introduced during the 1950s and early 1960s by Pierre Bezier and Paul de Casteljau [20] as mathematical tools for the CAGD systems developed by the French automobile companies Renault and Citroen.

Originally Bezier and de Casteljau introduced the Bezier curves from geometric considerations. Later, in 1970, Forrest [21] found the connection between the Bezier curves and the Bernstein polynomials. Thus, the notation introduced by Forrest will be used. A Bezier curve can be expressed in terms of the Bernstein basis as follows:

$$\vec{r}(t) = \sum_{i=0}^{n} \vec{b}_i B_i^n(t) \qquad 0 \leq t \leq 1 \tag{2.1}$$

where n is the curve degree and $\vec{b}_i$ are the control points, which are a set of points in the 3-D space that form the Bezier polygon (Figure 2.4).

The functions $B_i^n(t)$ are the Bernstein polynomials [8], which can be expressed as follows:

$$B_i^n(t) = \frac{n!}{i!(n-i)!} t^i (1-t)^{n-i} \tag{2.2}$$

Some important properties of the Bezier curves are:

- The number of control points that define the curve is equal to the curve degree plus one $(n+1)$.
- The curve domain is the segment in the parametric axis between 0 and 1.
- The Bezier curve follows approximately the shape of the Bezier polygon.

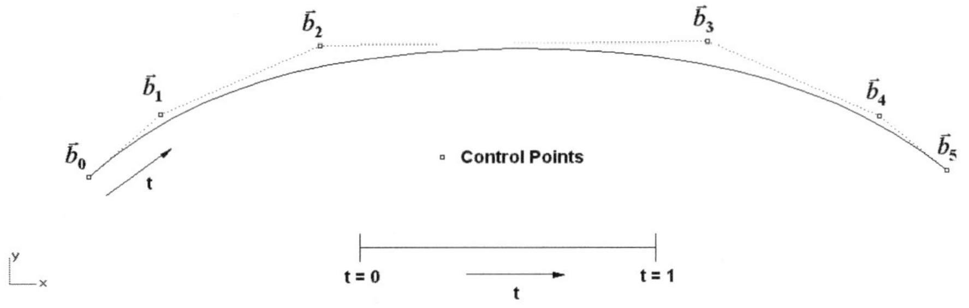

Figure 2.4 Bezier polygon defined by the control points used to represent a Bezier curve.

- Endpoint interpolation: the first and last control points ($\vec{b}_0$ and $\vec{b}_1$) coincide with the endpoints of the curve. This property can be easily proved by using the values $t = 0$ and $t = 1$ in (2.1) and (2.2).
- The vector that joins the first and second control points is tangent to the curve at the first endpoint. The vector that joins the penultimate and the last control points is tangent to the curve at the last endpoint.
- Convex hull property: the curve lies in the convex hull of the control points. The convex hull of a set of points is defined as the highest area polygon with vertices that are points of the original set of points.
- The number of intersections between an arbitrary plane and a Bezier curve is equal or less than the number of intersections between the plane and the corresponding Bezier polygon.
- The Bezier curves are invariant under affine transformations (scaling, translations, rotations).
- Linear precision: if the control points are located along a straight line, the Bezier curve is a segment contained in such a line.
- The Bezier curves are differentiable and all the derivatives are continuous functions.

In 1967, Coons introduced rational curves to CAGD. The expression of a rational Bezier curve [8, 22] is the following:

$$\vec{r}(t) = \frac{\sum_{i=0}^{n} w_i \vec{b}_i B_i^n(t)}{\sum_{i=0}^{n} w_i B_i^n(t)} = \frac{\vec{P}(t)}{W(t)} \tag{2.3}$$

where the coefficients w_i are the weights. Each control point has an associated weight. Working with rational Bezier curves has several advantages:

- They are able to represent conic curves (ellipses, hyperbolas, and parabolas) exactly. This cannot be done using nonrational curves.
- The weights provide the designer with an additional degree of freedom. When the weight of a control point increases, the curve is more attracted by the control point.
- The nonrational Bezier curves are a particular case of rational Bezier curves when all the weights are equal to one.
- In most CAGD systems the weights are greater than zero. When this condition is verified, the rational curves fulfill all the properties mentioned above for the nonrational Bezier curves.

The r-order parametric derivative of a rational Bezier curve can be expressed as follows:

$$\frac{d^r \vec{r}(t)}{dt^r} = \vec{r}_r(t) = \frac{\vec{P}_r(t) - \sum_{j=1}^{r} \binom{r}{j} W_j(t) \vec{r}_{r-j}(t)}{W(t)} \tag{2.4}$$

where

$$\vec{P}_r(t) = \frac{d^r \vec{P}(t)}{dt^r} = \frac{n!}{(n-r)!} \sum_{j=0}^{n-r} \Delta^r \vec{p}_j(t) B_j^{n-r}(t) \tag{2.5}$$

and

$$\Delta^r \vec{p}_j = \sum_{i=0}^{r} \binom{r}{i} (-1)^{r-i} w_{i+j} \vec{b}_{i+j} \tag{2.6}$$

The Bezier curves are a useful tool to design and represent curves, but they have some limitations:

- They do not verify the "local control" property. That is, a change in one of the control points affects the shape of the whole curve. Sometimes this is a problem for the designer.
- If the curve to be represented has a complex shape, it must be modeled using a high-degree Bezier curve. However, high degrees (8 or more) are considered to be inappropriate for practical applications because they are difficult to control (oscillations) and have a high computational cost.

To overcome these disadvantages, one can use composite Bezier curves known as splines. When, in the connections between the Bezier curves, certain continuity conditions (in the curve and/or in the parametric derivatives) are imposed, one obtains the B-spline curves. The B-spline curves overcome the disadvantages of the simple (noncomposite) Bezier curves, that is, complex curves can be modeled with low-degree curves and they verify the "local control" property. Currently, because of their flexibility, B-spline curves are the most widely used tool in CAGD systems for curve representation and modeling. Moreover, simple Bezier curves can be considered to be a particular case of B-spline curves.

2.3 RATIONAL B-SPLINE CURVES

The B-spline curves [8, 22, 23] were originally used by de Boor and other authors as a tool to solve curve approximation problems. Later, Gordon and Riesenfield introduced the use of the B-spline curves as a tool for curve design [24]. They also found the mathematical relation between Bezier and B-spline curves, proving that B-spline curves are a generalization of Bezier curves. Vesprille [25] was the first author to propose using rational B-spline curves. In 1996, Piegl and Tiller [26] established the mathematical basis of the rational B-spline curves.

A B-spline curve is a parametric polynomial curve that can be expressed as follows:

$$\vec{r}(t) = \sum_{i=1}^{n} \vec{d}_i N_i^k(t) \qquad 0 \le a \le t \le b, 2 \le k \le n+1 \tag{2.7}$$

where $N_i^k(t)$ are the B-spline basis, $k-1$ is the curve degree, and $\vec{d}_i$ are the n control points that form the B-spline polygon (or control polygon) associated with the curve (Figure 2.5). The basis N can be recursively defined as functions of a set of values in the curve parametric domain $\{t_i\}_{i=1}^{n-k}$, which are defined as part of the curve. These values are named knots and the set of knots constitutes the knot vector (or knot sequence). The basis functions can be written as follows:

$$N_i^1(t) = \begin{cases} 1 & \text{if } t_i \leq t \leq t_{i+1} \\ 0 & \text{otherwise} \end{cases} \tag{2.8}$$

$$N_i^k(t) = \frac{(t - t_i)N_i^{k-1}(t)}{t_{i+k-1} - t_i} + \frac{(t_{i+k} - t)N_{i+k}^{k-1}(t)}{t_{i+k} - t_{i+1}} \tag{2.9}$$

The knot vector is an increasing sequence of real numbers in the parametric domain of the curve. The multiplicity of a knot is defined as the number of times that its value appears in the knot vector. Usually, the multiplicity of the first and last knot of the knot sequence is equal to k; thus, the knot vector can be written as follows:

$$a = t_1 = t_2 = ... = t_k < t_{k+1} < t_{k+2} < ... < t_n < t_{n+1} = ... = t_{n+k} = b \tag{2.10}$$

In most references and CAGD systems, the curve parametric domain varies between 0 and 1; therefore, the first and last knots take the following values: $a = 0$ and $b = 1$. Hereafter, this assumption will be made for curves and surfaces.

From the above equations, one can conclude that $\vec{r}(t)$ is a composite function. Each component is a polynomial function of degree $k-1$ defined on the corresponding interval in the knot vector $t_i \leq t \leq t_{i+1}$.

The B-spline curves are totally differentiable with continuity in the whole domain except at the knot values. In a knot with multiplicity r, the curves are, at least, C^{k-1-r}; that is, the parametric derivatives of order $k - 1 - r$ and less are continuous in such a knot.

Depending on the knot vector, the curve can be classified as uniform or nonuniform. In the first case, the knot values are equally spaced in the parametric domain. In nonuniform curves, there is no such restriction. Usually, the CAGD systems use nonuniform curves because they provide more flexibility than the uniform ones.

Some important properties of the B-spline curves including the following:

- Local control property.
- The curve approximately follows the shape of the control polygon.

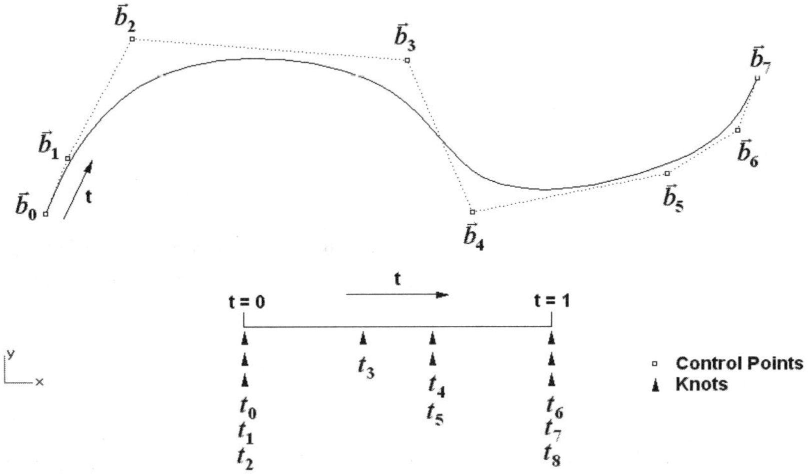

Figure 2.5 Control points and knot vectors that define a B-spline curve.

- Convex hull property.
- The B-spline curves are invariants under affine transformations (scaling, translations, rotations).
- The number of intersections between an arbitrary plane and a B-spline curve is equal to or less than the number of intersections between the plane and the corresponding control polygon.
- Linear precision.
- The number of knots equals the number of control points plus k.
- Endpoint interpolation.

In 1975, K. J. Vesprille introduced the rational B-spline curves as a generalization of the B-spline curves. When the rational B-spline curves are nonuniform, they are NURBS curves, which can be expressed as follows:

$$\vec{r}(t) = \frac{\sum_{i=1}^{n} w_i \vec{d}_i N_i^k(t)}{\sum_{i=1}^{n} w_i N_i^k(t)} \qquad 0 \leq a \leq t \leq b, 2 \leq k \leq n+1 \tag{2.11}$$

where the coefficients w_i are the weights. Each weight is associated with a corresponding control point. The usage of these curves has the following advantages:

- They are able to represent conic curves.
- They constitute a flexible tool for complex curve representation because of the weights and their composite nature.

2.4 RATIONAL BEZIER SURFACES

The rational Bezier surfaces [8, 22] were introduced in the late 1960s. Typically, they are defined as a generalization of the Bezier curves to two dimensions. The majority of their properties can be deduced from the Bezier curves' properties. The mathematical expression of a rational Bezier surface in terms of the Bernstein polynomials is:

$$\vec{r}(u,v) = \frac{\sum_{i=0}^{m}\sum_{j=0}^{n} w_{ij} \vec{b}_{ij} B_i^m(u) B_j^n(v)}{\sum_{i=0}^{m}\sum_{j=0}^{n} w_{ij} B_i^m(u) B_j^n(v)} = \frac{\vec{P}(u,v)}{W(u,v)} \tag{2.12}$$

where m and n are the degrees for the parametric coordinates u and v, respectively. Now, instead of a control polygon, the control points $\vec{b}_{ij}$ form a control mesh (Figure 2.6). Each control point has an associated weight w_{ij}. The number of control points for each parametric coordinate equals the degree plus 1. Therefore, the number of control points along the u coordinate is $m + 1$ and along the v coordinate is $n + 1$. Consequently, the total number of control points of the mesh is $(m + 1) \times (n + 1)$. Thus, the parametric space is a square in which both parametric coordinates vary between 0 and 1: $0 \leq u, v \leq 1$.

Some interesting properties of the Bezier surfaces are:

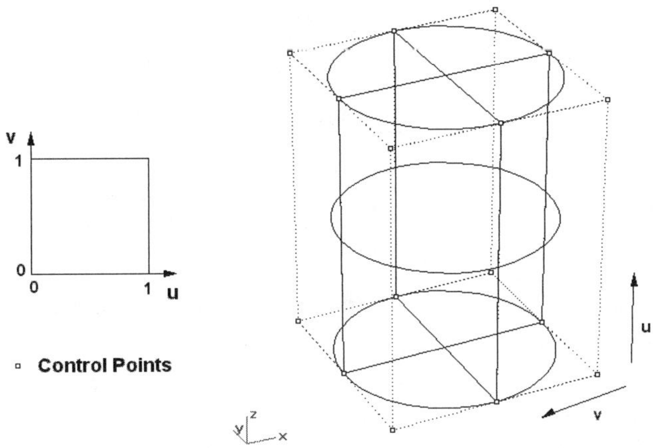

Figure 2.6 Control points that define a Bezier surface.

- The surface shape approximates the mesh of control points.
- Convex hull property.
- The isoparametric curves, $v = v_0 = constant$, are rational Bezier curves of degree m, which can be expressed as follows:

$$\vec{r}(u, v_0) = \frac{\sum_{i=0}^{m} w'_i \vec{b}'_i B_i^m(u)}{\sum_{i=0}^{m} w'_i B_i^m(u)} \tag{2.13}$$

where the control points $\vec{b}'_i$ and their associated weights w'_i are given by:

$$\vec{b}'_i = \frac{\sum_{j=0}^{n} w_{ij}\vec{b}_i B_{ij}^n(v_0)}{\sum_{j=0}^{n} w_{ij} B_j^n(v_0)} \tag{2.14}$$

$$w'_i = \sum_{j=0}^{n} w_{ij} B_{ij}^n(v_0) \tag{2.15}$$

Analogous reasoning can be used for the case of isoparametric curves, $u = u_0 = constant$.

- The isoparametric curves $v_0 = 0$ and $v_0 = 1$ coincide with the boundary curves of the surface and are defined by the corresponding control points (and weights) of the control mesh. For example, the isoparametric curves $v_0 = 0$ and $v_0 = 1$ are given by:

$$\vec{r}(u, 0) = \frac{\sum_{i=0}^{m} w_{i0}\vec{b}_{i0} B_i^m(u)}{\sum_{i=0}^{m} w_{i0} B_i^m(u)} \tag{2.16}$$

$$\vec{r}(u, 1) = \frac{\sum_{i=0}^{m} w_{in}\vec{b}_{in} B_i^m(u)}{\sum_{i=0}^{m} w_{in} B_i^m(u)} \tag{2.17}$$

Analogous reasoning can be used for the isoparametric curves $u_0 = 0$ and $u_0 = 1$.

- The parametric coordinates of the surface vertices are (0,0), (1,0), (1,1), and (0,1). They coincide with the four control points located in the vertices of the control mesh, that is, $\vec{b}_{00}$, $\vec{b}_{m0}$, $\vec{b}_{mn}$, and $\vec{b}_{mn}$, respectively.

- The Bezier surfaces are invariant under affine transformations (scaling, rotations, translations).

The parametric derivatives of the rational Bezier surfaces are given by:

$$\frac{\partial^{r+s}\vec{r}}{\partial u^r \partial v^s} = \frac{1}{W}\left[\vec{p}_{r,s} - \sum_{i=1}^{r}\sum_{j=1}^{s}\begin{pmatrix} r \\ i \end{pmatrix}\begin{pmatrix} s \\ j \end{pmatrix} W_{i,j}\vec{r}_{r-i,s-j} + \sum_{i=1}^{r}\begin{pmatrix} r \\ i \end{pmatrix} W_{i,0}\vec{r}_{r-i,s} - \sum_{j=1}^{s}\begin{pmatrix} s \\ j \end{pmatrix} W_{0,j}\vec{r}_{r,s-j}\right] \quad (2.18)$$

where

$$\vec{p}_{i,j} = \frac{m!n!}{(m-i)!(n-j)!}\sum_{k=0}^{m-i}\sum_{l=0}^{n-j}\Delta^{i,j}\vec{p}_{k,l}B_k^{m-i}(u)B_l^{n-j}(v) \quad (2.19)$$

$$W_{i,j} = \frac{m!n!}{(m-i)!(n-j)!}\sum_{k=0}^{m-i}\sum_{l=0}^{n-j}\Delta^{i,j}w_{k,l}B_k^{m-i}(u)B_l^{n-j}(v) \quad (2.20)$$

and the operator Δ is defined as follows:

$$\Delta^{i,j}\vec{p}_{k,l} = \Delta^{i-1,j}\vec{p}_{k+1,l} - \Delta^{i-1,j}\vec{p}_{k,l} \quad \text{when } i > 0 \quad (2.21)$$

$$\Delta^{i,j}\vec{p}_{k,l} = \Delta^{i,j-1}\vec{p}_{k,l+1} - \Delta^{i,j-1}\vec{p}_{k,l} \quad \text{when } j > 0 \quad (2.22)$$

$$\Delta^{i,j}w_{k,l} = \Delta^{i-1,j}w_{k+1,l} - \Delta^{i-1,j}w_{k,l} \quad \text{when } i > 0 \quad (2.23)$$

$$\Delta^{i,j}w_{k,l} = \Delta^{i,j-1}w_{k,l+1} - \Delta^{i,j-1}w_{k,l} \quad \text{when } j > 0 \quad (2.24)$$

$$\Delta^{0,0}\vec{p}_{k,l} = \vec{p}_{k,l} \qquad \Delta^{0,0}w_{k,l} = w_{k,l} \quad (2.25)$$

$$\vec{p}_{i,j} = \vec{b}_{i,j}w_{i,j} \quad (2.26)$$

The first-order parametric derivatives $\vec{r}_u$ and $\vec{r}_v$ are tangent vectors at the isoparametric curves $u = constant$ and $v = constant$, respectively.

The unit normal vectors at the surface points can be expressed as functions of the first-order parametric derivatives as follows:

$$\hat{n}(u,v) = \frac{\vec{r}_u \times \vec{r}_v}{|\vec{r}_u \times \vec{r}_v|} \tag{2.27}$$

Similarly, the infinitesimal area elements of the surface can be expressed as follows:

$$ds(u,v) = |\vec{r}_u \times \vec{r}_v|\, dudv \tag{2.28}$$

There are particular cases in which the functions $\hat{n}(u,v)$ and $ds(u,v)$ are not well defined in some of the surface vertices. This occurs when, at such points, $\vec{r}_u \times \vec{r}_v = 0$. In this case, the surface is said to degenerate at those points.

An important type of Bezier surface is the ruled surface [8]. The ruled surface is made from an infinite set of segments that join the points of two curves. These segments are a family of isoparametric curves. A rational Bezier surface is a ruled surface if it verifies one of the following two conditions:

$$\begin{aligned} n &= 1 \\ w_{i1} &= w_{i0}, i = 1, ..., m \end{aligned} \tag{2.29}$$

$$\begin{aligned} n &= 1 \\ w_{1j} &= w_{0j}, j = 1, ..., n \end{aligned} \tag{2.30}$$

If the first two conditions are fulfilled, the segments, which form the ruled surface, are isoparametric lines with $v = constant$, and the surface can be expressed as a linear function for the coordinate v:

$$\vec{r}(u,v) = \vec{r}_0(u) + v\vec{r}_1(u) \tag{2.31}$$

where

$$\vec{r}_0(u) = \frac{\sum_{i=0}^{m} w_{i0}\vec{b}_{i0}B_i^m(u)}{\sum_{i=0}^{m} w_{i0}B_i^m(u)} \tag{2.32}$$

$$\vec{r}_1(u) = \frac{\sum_{i=0}^{m} (w_{i1}\vec{b}_{i1} - w_{i0}\vec{b}_{i0}) B_i^m(u)}{\sum_{i=0}^{m} w_{i0} B_i^m(u)} \tag{2.33}$$

The function $\vec{r}_0(u)$ defines the isoparametric curve $v = 0$, and the function $\vec{r}_0(u, v) + \vec{r}_1(u, v)$ defines the isoparametric curve $v = 1$.

In the second case, the surface is linear for the coordinate u:

$$\vec{r}(u, v) = \vec{r}_0(v) + u\vec{r}_1(v) \tag{2.34}$$

where

$$\vec{r}_0(v) = \frac{\sum_{j=0}^{n} w_{0j}\vec{b}_{0j} B_j^n(v)}{\sum_{j=0}^{n} w_{0j} B_j^n(v)} \tag{2.35}$$

$$\vec{r}_1(v) = \frac{\sum_{j=0}^{n} (w_{1j}\vec{b}_{1j} - w_{0j}\vec{b}_{0j}) B_j^n(v)}{\sum_{j=0}^{n} w_{0j} B_j^n(v)} \tag{2.36}$$

The usage of Bezier surfaces has the following disadvantages:

- They do not follow the “local control” property.
- If the surface to be represented has a complex shape, it must be modeled using a high-degree Bezier surface. However, degrees higher than 8 are considered inappropriate for practical applications because of the oscillations introduced and the computational cost.

To overcome these disadvantages, as in the case of the curves, composite Bezier surfaces can be used. A set of Bezier curves joined under certain continuity conditions constitutes a B-spline surface. The B-spline surfaces overcome the disadvantages of the simple (noncomposite) Bezier surfaces; that is, complex surfaces can be modeled with low-degree surfaces and they verify the “local control” property. Because of their flexibility, B-spline surfaces are the most widely used tool in the CAGD systems for surface representation and modeling. Moreover, simple Bezier surfaces can be considered to be a particular case of B-spline surfaces.

2.5 RATIONAL B-SPLINE SURFACES

As Bezier patches, the B-spline surfaces [8, 22, 23, 27] can be introduced as a 2-D generalization of the rational B-spline curves. Therefore, they can be expressed as follows:

$$\vec{r}(u,v) = \frac{\sum_{i=0}^{m}\sum_{j=0}^{n} w_{ij}\vec{d}_{ij}N_i^k(u)N_j^l(v)}{\sum_{i=0}^{m}\sum_{j=0}^{n} w_{ij}N_i^k(u)N_j^l(v)} \qquad 2 \leq k \leq m+1, 2 \leq l \leq n+1 \quad (2.37)$$

where $k-1$ and $l-1$ are the surface degrees for the coordinates u and v, respectively; $\vec{d}_{ij}$ are the control points, which form the control mesh; w_{ij} are the weights; m and n are the number of control points for each parametric coordinate (the total number of control points will be mxn); and the functions N are the B-spline basis defined in Section 2.2. Figure 2.7 represents a typical rational B-spline surface.

As in the case of curves, most systems assume that the parametric coordinates vary between zero and one: $0 \leq u, v \leq 1$.

Now, there are two knot vectors, one for each parametric coordinate $\{u_i\}_{i=1}^{m+k}$ and $\{v_i\}_{i=1}^{n+l}$:

$$0 = u_1 = u_2 = ... = u_k < u_{k+1} < u_{k+2} < ... < u_m < u_{m+1} = ... = u_{m+k} = 1 \quad (2.38)$$

$$0 = v_1 = v_2 = ... = v_k < v_{k+1} < v_{k+2} < ... < v_n < v_{n+1} = ... = v_{n+k} = 1 \quad (2.39)$$

If, in both knot vectors, the knots are equally spaced, the surface is said to be uniform. Otherwise, the surface is a NURBS surface. Uniformity constitutes a particular case of NURBS surfaces.

Some important properties of these surfaces are the following:

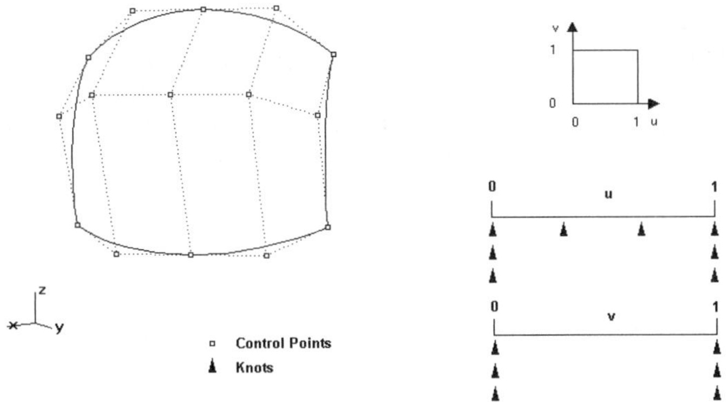

Figure 2.7 Control points and knot vectors that define a B-spline surface.

- They are invariant under affine transformations (scaling, rotations, translations).
- In the surface points, there is C^{k-2} continuity with respect to the parametric coordinate u and C^{l-2} with respect to the parametric coordinate v.
- Convex hull property.
- Local control property: a change in one of the control points affects only the part of the surface where the control point has influence.
- The four boundary curves of the surface are B-spline curves. Their control points are the boundaries of the control mesh, and their degrees and knot vectors are the same as the corresponding parametric coordinate.

2.6 TRANSFORMATION FROM NURBS REPRESENTATION TO BEZIER REPRESENTATION

In 1972, Carl de Boor proposed his "knot insertion" algorithm to calculate points in B-spline curves [28]. Using this technique, a new knot in the knot vector of the curve can be introduced (or the multiplicity of an existing knot can be increased) without changes in the shape of the curve. This process is known as knot insertion. When a knot is inserted in the knot vector of a curve, $k - 1$ of the original control points are replaced by k new control points, where $k - 1$ is the degree of the curve. More generally, when a knot is inserted s times, $k - 1$ old control points are replaced by $k-1+s$ new control points. When a knot is inserted repeatedly until its multiplicity equals the curve degree $(k - 1)$, the control point that arises in the last insertion coincides with the Cartesian coordinate of the curve at the knot. In other words, if a knot multiplicity equals the curve degree, the curve passes on the corresponding control point, and the parametric coordinate of the curve at such a point is the knot value.

In 1981, based on the work of de Boor, W. Boehm developed an algorithm to transform the B-spline description of a curve to a description in terms of Bezier curves [19]. This algorithm is based on the following fact: if the multiplicity of all the knots of a B-spline equals the curve degree, the control points coincide with the control points of a set of Bezier curves (composite Bezier curve). The union of these Bezier curves describes the curve exactly. The degree of the resulting Bezier curves coincides with the degree of the original B-spline curve. Each one of the Bezier curves is associated with an interval of the knot vector. Therefore, the number of the resulting Bezier curves equals the number of intervals in the knot vector of the B-spline description. Figure 2.8 shows an example of the transformation of a NURBS curve into Bezier curves.

Consequently, to obtain the Bezier description of a B-spline curve, new knots must be inserted on the original knots until their multiplicities equal the curve degree.

The algorithm proposed by W. Boehm can be easily generalized to surfaces. In this case, there are two knot vectors (one for each parametric coordinate). After the knot insertion process, the resulting control points coincide with the control points of a set of Bezier surfaces (composite Bezier surface) that describe the surface exactly. The number of resulting Bezier surfaces equals the product of the number of intervals in the knot vectors. Figure 2.9 shows an example of the transformation of a NURBS surface into Bezier patches.

W. Boehm named this method the "Cox de Boor" algorithm. There are other algorithms to obtain the Bezier patches representation of NURBS surfaces, but, to the best of our knowledge, the Cox de Boor algorithm is the most efficient.

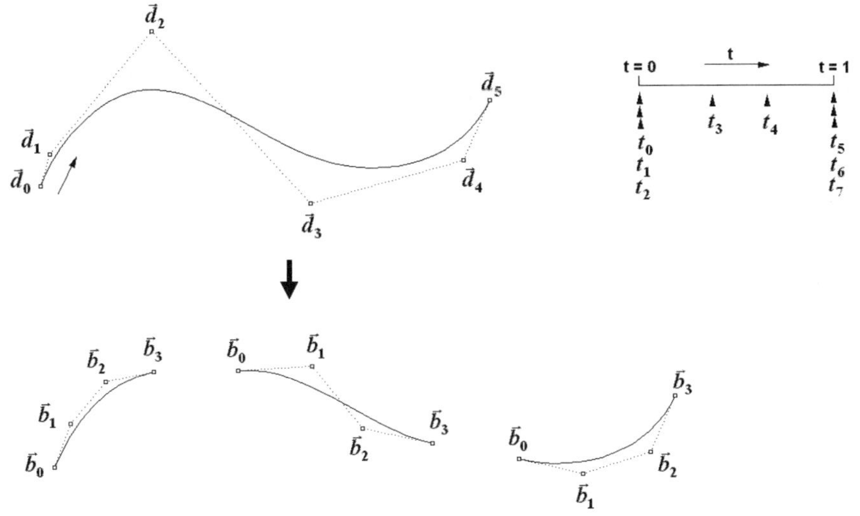

Figure 2.8 Transformation of a B-spline curve in Bezier curves.

2.7 COMPUTATION OF GEOMETRIC PARAMETERS ON BEZIER SURFACES

2.7.1 Computation of Unit Normal Vectors

The unit normal vectors at a point over a Bezier patch can be expressed as functions of the first-order parametric derivatives as follows:

$$\hat{n}(u,v) = \frac{\vec{r}_u \times \vec{r}_v}{|\vec{r}_u \times \vec{r}_v|} \tag{2.40}$$

where $\vec{r}_u$ is the first-order derivative with respect to the coordinate u, and $\vec{r}_v$ is the first-order derivative with respect to the coordinate v. Such derivatives can be calculated using (2.18) when $r = 1, s = 0$ and $r = 0, s = 1$, respectively.

The normal vector defines the "active" side of the surface; that is, the side of the surface viewed from points outside the body.

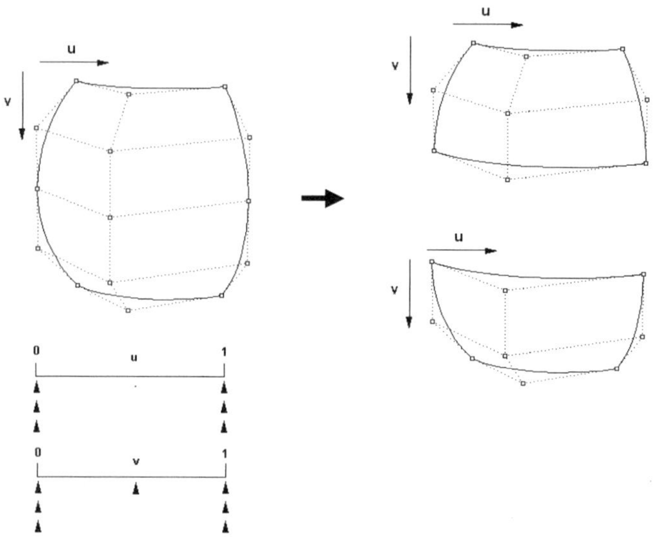

Figure 2.9 Transformation of a B-spline surface into Bezier surfaces.

2.7.2 Computation of Principal Curvature Radii and Principal Directions

Considering a point on a surface, and one of the infinite planes containing the point and the normal vector at the point, the intersection of the surface and the plane defines a line on the surface that contains the point. This line has a curvature at that point. If another plane is considered, another line with a different curvature at that point is defined. The maximum and minimum of the infinite set of curvatures (corresponding of the infinite possible planes) are the principal curvatures, and the corresponding planes are the principal planes. The principal directions are defined as the intersections between the principal planes and the tangent plane to the surface at the point. Clearly, the principal directions are vectors that are perpendicular to the normal vectors and tangent to the surface at the point. Moreover, the principal directions are perpendicular to each other. The principal directions have a simple geometric meaning: they indicate the directions of maximum and minimum curvature of the surface at the point.

It is important to remark that the principal curvatures and principal directions are functions of the surface points. In other words, they are functions of the surface parametric coordinates. Therefore, the principal curvatures and principal directions vary on the surface. The principal curvatures are scalar functions, and the principal dircctions are vector functions.

The principal curvatures and principal directions can be obtained in terms of the parameters of the first and second fundamental forms of differential geometry. These parameters can be calculated from the parametric derivatives of the surface. The parameters of the first fundamental form are

$$E = \vec{r}_u \cdot \vec{r}_u, \qquad F = \vec{r}_u \cdot \vec{r}_v, \qquad G = \vec{r}_v \cdot \vec{r}_v \tag{2.41}$$

and the parameters of the second fundamental form are

$$L = \hat{n} \cdot \vec{r}_{uu}, \qquad M = \hat{n} \cdot \vec{r}_{uv}, \qquad N = \hat{n} \cdot \vec{r}_{vv} \tag{2.42}$$

where $\vec{r}_u$, $\vec{r}_v$, $\vec{r}_{uu}$, $\vec{r}_{uv}$, and $\vec{r}_{vv}$ are the first- and second-order parametric derivatives, and $\hat{n}$ is the unit normal vector.

The principal curvatures are the roots of the following second-order polynomial

$$k^2(EF - F^2) + k(2FM - EN - LG) + (LN - M^2) = 0 \tag{2.43}$$

It can be shown that the roots of the expression (2.43) are always real numbers. The sign of the roots indicates whether the surface is convex or concave. Looking at the active side of the surface, the surface is convex when the principal curvatures are positive, and the surface is concave when the principal curvatures are negative.

In ruled surfaces, one of the principal curvatures is zero at all points. Typical examples are cylinders and conical surfaces. When both principal curvatures are zero at all points, the surface is a plane. The points where one curvature is positive and the other negative are called saddle points. At such points, the surface is concave and convex.

When $L : M : N = E : F : G$, the principal curvatures coincide; thus, the curvatures are constant for the different planes. These points are the umbilical points. Typical examples are the points of a sphere. The points of a flat surface can be viewed as umbilical points with zero curvature.

If the surface is viewed from the nonactive side, the curvatures can be obtained by changing the sign of the principal curvatures viewed from the active side. In other words, when the sign of the normal vector changes, the signs of the principal curvatures change.

The principal directions can be calculated with the parameter λ. Its value defines a direction in the parametric space of the surface:

$$\lambda = \frac{dv}{du} \tag{2.44}$$

and it also defines a direction in the tangent plane to the surface at the point. Therefore, the principal directions are associated with two specific values of the parameter λ. These values are the roots of the following polynomial:

$$\lambda^2(FN - GM) + \lambda(EN - LG) + (EM - FL) = 0 \tag{2.45}$$

It can be shown that the roots of expression (2.45) are always real numbers. Let us suppose that λ_1 , λ_2 are the roots. Then the principal directions $\vec{d_1}$ and $\vec{d_2}$ are given by

$$\vec{d_1} = \lambda_1 \vec{r}_u + \vec{r}_v, \qquad \vec{d_2} = \lambda_2 \vec{r}_u + \vec{r}_v \tag{2.46}$$

References

[1] J.W. Crispin and A.L. Maffet. "Radar cross section estimation for complex shapes." *Proceedings of the IEEE*, 53(8):972–983, August 1965.

[2] D. Meagher. "Geometric modeling using octree encoding." *Computer Graphics and Image Processing*, 19(2):129–147, June 1982.

[3] M. E. Mortenson. *Geometric Modeling*. John Wiley and Sons, 1985.

[4] M. Mantyla and R. Sulonen. "GWB: A solid modeler with Euler operators." *IEEE Computer Graphics and Applications*, 2(7):17–31, September 1982.

[5] A. G. Requischa and H. B. Voelcker. "Solid modeling: Current status and research directions." *IEEE Computer Graphics and Applications*, 3(7):25–37, October 1983.

[6] N. N. Youssef. "Radar cross section of complex targets." *Proceedings of the IEEE*, 77(5):722–734, May 1989.

[7] M. Domingo, R. P. Torres, and M. F. Cátedra. "Calculation of the RCS from the interaction of edges and facets." *IEEE Transactions on Antennas and Propagation*, 42(6):885–888, June 1994.

[8] G. Farin. *Curves and Surfaces for Computer Aided Geometric Design*. Academic Press, 1988.

[9] W. Dahmen, M. Gasca, and C. A. Micchelli. *Computation of Curves and Surfaces*. Kluwer Academic Publishers, 1990.

[10] J. Pérez and M. F. Cátedra. "Application of physical optics to the RCS computation of bodies modeled with NURBS surfaces." *IEEE Transactions Antennas and Propagation*, 42(10):1404–1411, October 1994.

[11] J. Pérez, J. A. Saiz, O. M. Conde, R. P. Torres, and M. F. Cátedra. "Analysis of antennas on board arbitrary structures modeled by NURBS surfaces." *IEEE Transactions on Antennas and Propagation*, 45(6):1045–1053, June 1997.

[12] J. L. Hu, S. M. Lin, and W. B. Wang. "Computation on PO integral on NURBS surface and its application to RCS calculation." *Electronic Letters*, 33(3):239–240, January 1997.

[13] M. Chen, Y. Zhan, X. W. Zhao, and C. H. Liang. "Analysis of antenna around NURBS surface with hybrid MoM-PO technique." *IEEE Transactions on Antennas and Propagation*, 55(2):407–413, February 2007.

[14] Y. C. Ruan, X. Y. Zhou, J. Y. Chin, T. J. Cui, Y. B. Tao, and H. Lin. "The UTD analysis to EM scattering by arbitrarily convex objects using ray-tracing of creeping waves on numerical meshes." *IEEE Antennas and Propagation International Symposium*, pp. 1–4, July 2008.

[15] H. A. Serim and A. A. Ergin. "Computation of the physical optics integral on NURBS surfaces using a radon transform interpretation." *IEEE Antennas and Wireless Propagation Letters*, 7:70–73, 2008.

[16] Web page of Microstation: www.bentley.com

[17] Web page of Rhinoceros: www.rhino3d.com

[18] IGES/PDES Organization National Computer Association, administrator. "The initial graphics exchange specification (IGES) version 5.0."

[19] W. Boehm. "Generating the Bezier points of B-spline curves and surfaces." *Computer Aided Design*, 13:365–366, November 1981.

[20] P. de Casteljau. *Mathématiques et CAO. Vol. 2 : Formes á pôles*. Hermes, 1986.

[21] A. R. Forrest. "Interactive interpolation and approximation by Bezier polynomials." *The Computer Journal*, 15(1):71, 79 1972.

[22] D. F. Rogers and J. A. Adams. *Mathematical Elements for Computer Graphics*. McGraw-Hill, 1989.

[23] W. Tiller. "Rational B-splines for curve and surface representation." *IEEE Computer Graphics and Applications*, 3(6):61–69, September 1983.

[24] W. J. Gordon and R. F. Riesenfeld. "B-spline curves and surfaces." *Computer Aided Geometric Design*, pp. 95–126. Academic Press, 1974.

[25] K. J. Versprille. *Computer-Aided Design Applications of the Rational B-Spline Approximation Form*. Ph.D. Thesis, Syracuse University, May 1975.

[26] L. Piegl and W. Tiller. *The NURBS Book*. Springer-Verlag, 1996.

[27] W. Boehm and G. Farin. "A survey of curve and surface methods in CAGD." *Computer Aided Geometric Design*, 1(1):1–60, July 1984.

[28] C. de Boor. *A Practical Guide to Splines*. Springer-Verlag, 1978.

Chapter 3

Geometrical Optics and Uniform Theory of Diffraction (GO/UTD)

3.1 HISTORICAL OVERVIEW

The study of light phenomenon originated with the Greeks when they tried to find a mathematical and geometrical explanation to its propagation using observation [1]. In the sixteenth century and at the beginning of the seventeenth century, the field of the optics expanded greatly with the development of optical instruments (Kepler, Galileo) and significant discoveries, such as the refraction law (Snell) and Fermat's principle [2]. Based on these theories, all the mathematics behind optics were established and are called classical geometrical optics (GO) [3]. This theory did not take into account the electromagnetic character of light, and it did not include such concepts as the phase, polarization, and amplitude in the study of the propagation.

The appearance of Maxwell's equations in 1864 marks a clear evolution in electromagnetic theory, and its results are applicable to wave propagation and consider light to be a high-frequency electromagnetic wave. Sommerfield and Range [4], based on mathematical and physical concepts derived from Poincaré's work [5], were the pioneers who linked Maxwell's equations with the GO laws by introducing the notion of an asymptotic series. The extension of this work to vector fields by Luneberg [6] and Kline [7] led to the modern version GO, which proves that GO expressions correspond to the first term of an asymptotic series.

However, GO presented the problem of shadow areas created by edges and discontinuities (Figure 3.1) because it was unable to predict the field in these areas. This problem was solved by Keller [8, 9] in 1953 by introducing diffraction, which laid the foundations of the geometrical theory of diffraction (GTD). Keller introduced diffracted rays, which together with GO's direct and reflected rays, allowed the high-frequency field to be predicted in the presence of objects with edges. Keller stated that the diffracted rays behave as GO rays once they leave the edge, and their path can be obtained by generalizing Fermat's principle [2]. This principle states that the path followed by a ray between two points (source and observer) is always such that the optical path length is an extremum (maximum or minimum). In a homogeneous medium, such as the air, the optical length is equal that the geometric length; thus, the ray trajectories are straight lines.

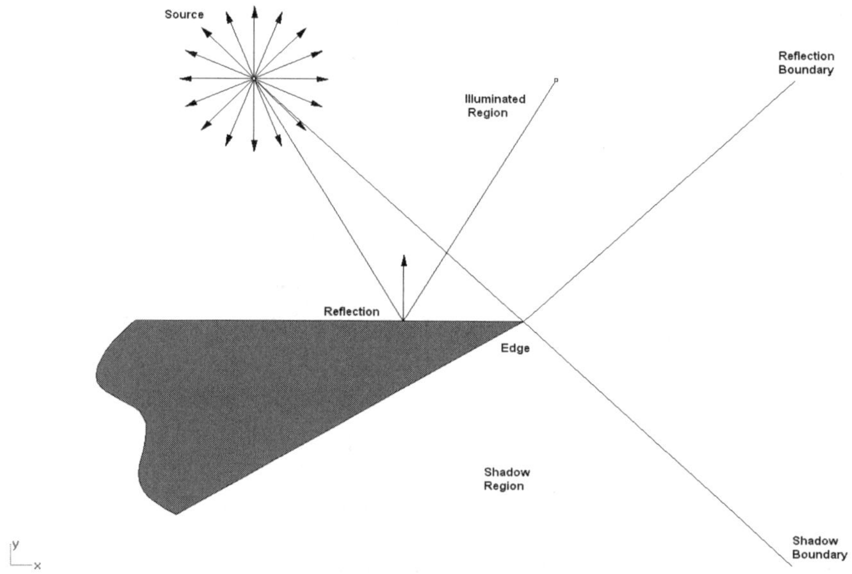

Figure 3.1 Example of the different regions defined by GO rays.

The original GTD formulation, developed by Keller, presented some problems, especially in the shadow and reflection boundaries where the GO fields present discontinuities. Additionally, the diffracted field, according to Keller's formulation, becomes infinite because this formulation is only valid outside these boundaries. These problems were solved with uniform solutions to the asymptotic expressions that Keller used to develop his theory. Two solutions were introduced for edge diffraction. The first one is the uniform theory of diffraction (UTD) developed by Kouyoumjian and Pathak [10] from the asymptotic expansion by Pauli Clemmow [11], while the second one is the uniform asymptotic solution (UAT) obtained by Lee and Deschamps [12] from the expansion by Oberhettinger and Van der Waerden [13]. The main advantage of the UTD is that it is numerically easier to implement compared to the UAT because the UAT obtains the fields by subtracting two terms whose values are infinite at the shadow boundaries. Thus, the UTD has been, historically, more widely used to solve realistic engineering problems, and therefore, it will be the focus of this chapter.

One of the applications in which UTD has been more widely used is analyzing on-board antennas in complex structures, such as aircrafts, satellites, and ships. The origins of this application can be dated to 1973 when W. D. Burnside, R. J. Marhefka, and C. L. Yu [14] applied the UTD formulation to analyze an antenna on-board an aircraft. The aircraft was modeled in this work using canonical structures, such as cylinders, to represent the fuselage and plane facets to represent the wings. Therefore, the model of the aircraft was not very complicated, which allowed the reflection and diffraction points to be obtained easily. However, the main limitation of this model was that the cylinders had infinite length, and therefore the results were not correct for certain observation directions.

In 1975, Burnside [15] proposed a new approach to obtain results of the radiation pattern of antennas on-board aircrafts, using a 1/25 scaled model of a KC-135. The idea was to increase the angular margin where the results can be considered good enough with respect to Burnside's previous paper. This objective is accomplished using a different geometric model to represent the aircraft depending on the cut of the desired radiation pattern and the position of the antenna. Two models were used: plane facets for the wings and an infinite-length cylinder with an elliptical section where the antenna is over the fuselage and a cut in the azimuth direction is analyzed, and the junction of two semiellipses where the cut is in elevation. However, the results were not correct for certain directions because some parts of the aircraft were not considered in this kind of model. Burnside and Gilreath [16] improved the model by including a vertical stabilizer modeled by plane facets, which improved the accuracy of the results. This model was also used to obtain results in the near field [17]. Further efforts mainly focused on modeling the aircraft more accurately. The works by Balanis and Cheng [18] and Kim and Burnside [19] can be mentioned, although they always solved the problem using canonical structures.

As can be seen, the main problem in obtaining accurate results is caused by the geometric representation of the complex object. When parametric surfaces, such as NURBS surfaces, appear, the problem of geometric modeling is solved. As mentioned in Chapter 2, NURBS surfaces allow complex models to be represented with a small amount of information and high accuracy. Thus, it is applicable to electromagnetic problems and, specifically, to the analysis of on-board antennas using GO/UTD. The foundation paper of this application [20] uses NURBS surfaces for the model that was previously used for the radar cross section (RCS) computation [21]. Since then, several authors have applied similar procedures to solve the problem of antennas on-board aircrafts by improving the ray-tracing formulation [22–24] (this aspect will be covered more in depth in Chapter 5) or by including some contributions to the radiated field that were not included previously, such as the creeping waves formulation [25, 26].

GO/UTD has not only been applied to the analysis of antennas on-board aircrafts; other structures, such as satellites [27, 28] or ships [29–31], have also been treated using this formulation. Moreover, GO/UTD has been widely used to predict mobile communication propagation for both urban [32–35] and indoor [36–39] environments. In this kind of problem, given the morphological structure of the environment, plane facet models have been preferred. The application of GO/UTD to plane facet modeling for mobile communications problems has been covered elsewhere [40], and it will not be repeated here because it is outside the scope of this book. However, even in environments where the plane facets are predominant, the combination of plane facet modeling with some structures modeled by NURBS can be advisable. There are some structures with arbitrary shapes in satellites and ships, but also (especially given current architectural tendencies) in cities and buildings. Therefore, NURBS modeling can be useful in this kind of application [41, 42] and is sometimes combined with parts modeled by plane facets. This aspect will be discussed in Chapter 6, which is devoted to the presentation of realistic cases in which asymptotic techniques combined with NURBS representation have been applied.

With all the considerations mentioned above, this chapter covers the application of GO/UTD to the analysis of electromagnetic problems over complex bodies modelled by parametric surfaces. The different ray contributions are treated while considering classical GO rays (direct and reflected), diffraction, and the phenomenon of creeping rays. The sections devoted to any contribution are divided into two parts: the electromagnetic formulation, which is obviously fundamental for the GO/UTD application, and, equally important, the procedure to perform the ray-tracing where parametric surfaces are involved. This part is the main contribution of this book, with the aim, as mentioned before, of providing engineers with a tool to solve realistic problems.

3.2 GEOMETRICAL OPTICS EQUATIONS

Initially, the field propagation according to GO was developed intuitively, without clear theoretical support, until it was necessary to include the electromagnetic nature in the phenomenon using Maxwell's equations. Starting from an isotropic, linear, lossless medium without sources, with electrical permittivity ϵ and magnetic permeability $\mu = \mu_0$, Maxwell's equations in the frequency domain can be written as [43]:

$$\nabla \times \vec{E}(\vec{r},\omega) + j\omega\mu_0\vec{H}(\vec{r},\omega) = 0 \tag{3.1}$$

$$\nabla \times \vec{H}(\vec{r},\omega) - j\omega\epsilon\vec{E}(\vec{r},\omega) = 0 \tag{3.2}$$

$$\nabla \cdot \vec{E}(\vec{r},\omega) = 0 \tag{3.3}$$

$$\nabla \cdot \vec{H}(\vec{r},\omega) = 0 \tag{3.4}$$

where $\vec{E}(\vec{r},\omega)$ and $\vec{H}(\vec{r},\omega)$ are the electric and magnetic field vectors at any point $\vec{r}$ of the space, respectively. By eliminating one of the vectors of the equations given above, the Helmholtz equation is obtained. This equation must be satisfied by any electromagnetic field that is propagated along the medium mentioned above:

$$\nabla^2 \cdot \vec{E}(\vec{r},\omega) + k^2\vec{E}(\vec{r},\omega) = 0 \tag{3.5}$$

where $k^2 = \omega^2\mu_0\epsilon$, and considering the speed of light in the medium, it can be written as $k = 2\pi/\lambda$, or the wave number.

Luneberg and Kline [6, 7] developed the formulation for electromagnetic fields at high frequency for an isotropic medium without sources as an infinite series of functions as follows:

$$\vec{E}(\vec{r},\omega) \sim e^{-jk\Psi(\vec{r})} \sum_{n=0}^{\infty} \frac{\vec{E}_n(\vec{r})}{(j\omega)^n} \tag{3.6}$$

$$\vec{H}(\vec{r},\omega) \sim e^{-jk\Psi(\vec{r})} \sum_{n=0}^{\infty} \frac{\vec{H}_n(\vec{r})}{(j\omega)^n} \tag{3.7}$$

where k is the wave number and $\Psi(\vec{r})$ is the phase or eikonal function [44]. It is important to note that the electric and magnetic fields, or each series and phase function, are independent from the frequency w, and they only depend on the space coordinate $\vec{r}$. When the frequency tends toward infinity, only the first term ($n = 0$) of the series remains, and it is called the geometrical optics field:

$$\vec{E}(\vec{r},\omega) \sim \vec{E}_0(\vec{r})e^{-jk\Psi(\vec{r})} \tag{3.8}$$

$$\vec{H}(\vec{r},\omega) \sim \vec{H}_0(\vec{r})e^{-jk\Psi(\vec{r})} \tag{3.9}$$

From these expressions, it is easy to obtain the power flow density [44] and the direction of propagation of the power $\hat{s}$ that defines the direction of the GO ray:

$$\hat{s} = (\nabla\Psi(\vec{r})) \tag{3.10}$$

It can be seen that the GO rays are tangent curves to the direction of $\hat{s}$, and energy transportation is carried out along these trajectories. Therefore, no energy is transported transversally to a ray. The ray tube is formed by considering the set of rays that transport enough energy, as shown in Figure 3.2. This ray tube will be used to determine the trajectory, phase, amplitude, and polarization along the ray propagation.

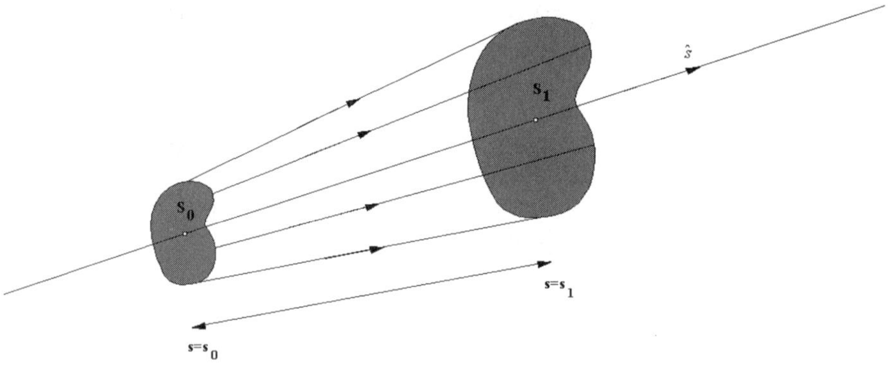

Figure 3.2 Ray tube.

3.2.1 Ray Trajectory

It can be proved [3, 44] that the GO rays in a homogeneous medium obtained from expressions (3.8) and (3.9) are straight lines. This does not happen in a nonhomogeneous medium in which the trajectories are usually curved.

3.2.2 Polarization

Polarization is defined as the orientation of the electric field vector at a point in space. It is represented as the unit vector $\hat{e}$, which is defined as:

$$\hat{e} = \frac{\vec{E}_0}{|\vec{E}_0|} \tag{3.11}$$

It is easy to prove that the polarization vector is perpendicular to the propagation direction $\hat{s}$ and satisfies the condition:

$$\hat{e} \cdot \hat{s} = 0 \tag{3.12}$$

The variation of the field polarization along the ray trajectory can also be obtained. In the case of a homogeneous medium, the polarization vector is constant:

$$\frac{d\hat{e}}{ds} = 0 \tag{3.13}$$

However, as will be shown, the polarization can be modified after a reflection or diffraction, or remain constant once the ray propagates from the obstacle where the reflection or the diffraction occurs.

3.2.3 Phase

The phase term of the fields satisfies the eikonal equation:

$$|\nabla\Psi(\vec{r})|^2 = 0 \tag{3.14}$$

By solving this equation, three important wavefronts can be obtained: plane, cylindrical, and spherical.

1. Plane wavefront: the surface defining the wavefront is a plane in which the amplitude and the phase of the fields remain constant. This phenomenon is typical when the source is very far from the observer.

$$\Psi(x, y, z) = ax + by + cz \tag{3.15}$$

2. Cylindrical wavefront: in this case, the surface defining the wavefront is a cylindrical section, which corresponds, in general, with a source that is a current line.

$$\Psi(x, y, z) = \sqrt{x^2 + y^2} \tag{3.16}$$

3. Spherical wavefront: the surface defining the wavefront is a spherical section and corresponds with the point sources, which occurs in radiation and propagation problems.

$$\Psi(x, y, z) = \sqrt{x^2 + y^2 + z^2} \tag{3.17}$$

In isotropic media, the rays are straight, and the phase is a constant over the surface, called the wavefront. Thus [45]:

$$d\Psi(\vec{r}) = |\nabla\Psi(\vec{r})|ds = ds \tag{3.18}$$

where ds is the distance increment along the ray trajectory. As a result, the difference from a reference point at $s = s_0$ to a point that is a distance s from the reference point is:

$$\Psi(s) - \Psi(s_0) = s - s_0 \tag{3.19}$$

If the reference point is at the point $s_0 = 0$, the function at the point s is

$$\Psi(s) = \Psi(0) + s \tag{3.20}$$

Therefore, the phase variation along the ray path is given by

$$e^{-jk\Psi(s)} = e^{-jk\Psi(0)}e^{-jks} \tag{3.21}$$

3.2.4 Amplitude

The last feature to determine is the amplitude of the fields along the propagation trajectory. It is only necessary to consider an infinitesimally narrow ray tube over a central ray that starts from a reference point at $s = 0$ and travels to another point placed at a distance s with propagation direction $\hat{s}$ and two closed phase surfaces centered at both points, as shown in Figure 3.3. This is an astigmatic ray tube. The phase surface with a reference at $s = 0$ is defined by the two principal radii of curvature ρ_1 and ρ_2, which are measured at the central ray [43].

Because the energy conservation theorem must hold along the ray tube, the electric field at a distance s can be obtained by relating areas of the phase surfaces, given the expression called divergence factor:

$$\sqrt{\frac{dS_0}{dS}} = \sqrt{\frac{\rho_1\rho_2}{(\rho_1 + s)(\rho_2 + s)}} \tag{3.22}$$

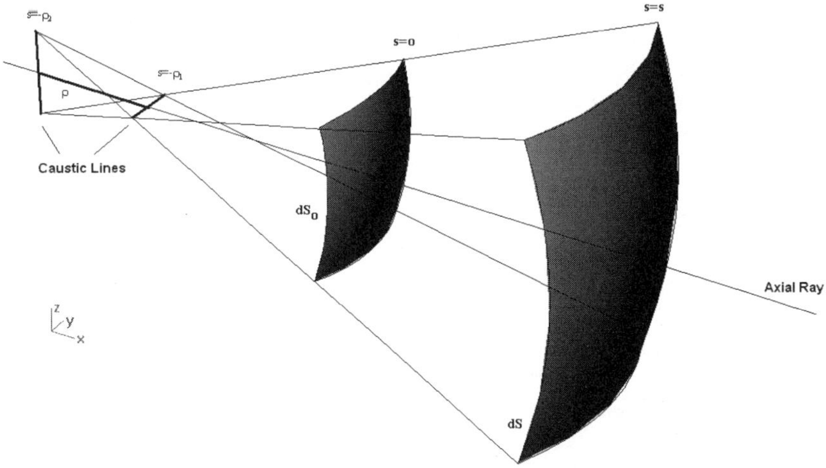

Figure 3.3 Astigmatic ray tube.

Therefore, the electric field, according to GO for an observation point placed at a distance s from a reference point placed at $s = 0$, can be written as [46]

$$\vec{E}(s) = \vec{E}(s = 0)\sqrt{\frac{\rho_1 \rho_2}{(\rho_1 + s)(\rho_2 + s)}} e^{-jks} \tag{3.23}$$

where $\vec{E}(s = 0)$ is the electric field at the reference point or phase center and is given by

$$\vec{E}(s = 0) = \vec{E}_0 e^{-jk\Psi(0)} \tag{3.24}$$

It can be observed from (3.23) that when $s = -\rho_1$ or $s = -\rho_2$, the field given by the GO is infinite, which is not correct. Thus, there are infinite rays going through the observation point; this phenomenon is referred to as caustic.

Because of the linearity of the previous equations, the total electric field at an observation point is the sum of all the rays that reach this point [40]

$$\vec{E}_T(s) = \sum_{i=0}^{N} \vec{E}_i \tag{3.25}$$

where N is the number of rays that reach the point directly, as well as after reflection on a surface, diffraction on an edge or combinations of the previous effects. Figure 3.4 shows an example for the position of a source and an observation point.

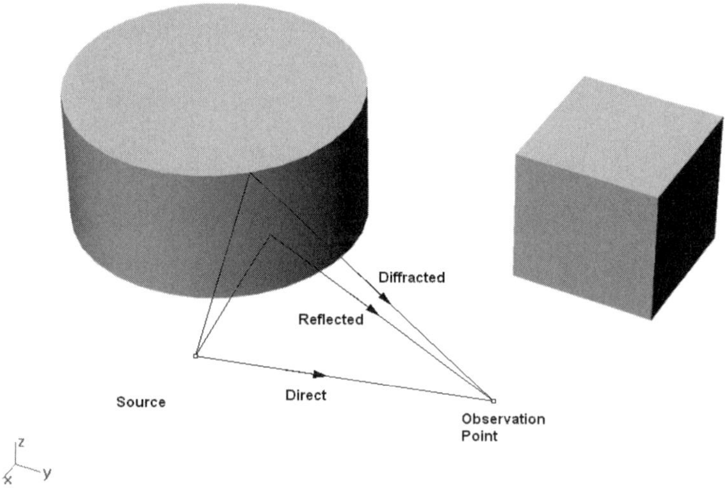

Figure 3.4 Example of rays that reach an observation point.

3.3 DIRECT RAY

3.3.1 Electromagnetic Field Calculation

The direct ray is the ray that proceeds directly from the source to the observation point. The electric field associated with the direct ray is the direct field. Expression (3.23) can be used to evaluate this electric field. For a point source, the wavefront is spherical $\rho_1 = \rho_2 = \rho$, and the amplitude factor then reduces to $A(S) = \rho/(\rho + s)$. Referring the field to the source point, expression (3.23) can be rewritten as follows:

$$\vec{E}(s) = \vec{E}(0)\frac{1}{s}e^{-jks} \tag{3.26}$$

3.3.2 Determination of the Direct Rays

To determine if there is a direct ray in a given direction (far-field analysis) or toward a given point (near-field analysis), it is necessary to check if there is any NURBS surface of the model that occludes the ray. If occlusion occurs, there is no direct ray in that direction. If the ray intersects none of the surfaces, there is a direct ray.

Therefore, to determine the direct rays, the ray-surface intersection test is made repeatedly. An efficient procedure is described here. A function $d(u, v)$ is considered. This function is defined as the distance between a ray and a point $\vec{r}(u, v)$ in a NURBS surface:

$$d^2(u, v) = (\vec{r}(u, v) - \vec{S}) \cdot (\vec{r}(u, v) - \vec{S}) - [\hat{v} \cdot (\vec{r}(u, v) - \vec{S})]^2 \tag{3.27}$$

where $\vec{S}$ and $\hat{v}$ are the origin and the direction of the ray, respectively. The partial derivatives of $d^2(u, v)$ are given by:

$$\frac{\partial}{\partial u}[d^2(u, v)] = 2\vec{s}(u, v) \cdot \vec{r}_u(u, v) - 2(\hat{v} \cdot \vec{s}(u, v)) \cdot (\hat{v} \cdot \vec{r}_u(u, v)) \tag{3.28}$$

$$\frac{\partial}{\partial v}[d^2(u, v)] = 2\vec{s}(u, v) \cdot \vec{r}_v(u, v) - 2(\hat{v} \cdot \vec{s}(u, v)) \cdot (\hat{v} \cdot \vec{r}_v(u, v)) \tag{3.29}$$

where $\vec{s}(u, v)$ is defined as follows:

$$\vec{s}(u, v) = \vec{r}(u, v) - \vec{S} \tag{3.30}$$

Minimizing $d^2(u, v)$, we obtain the surface point closest to the ray (u_0, v_0) and the shortest distance between the ray and the surface $d(u_0, v_0)$. To minimize (3.27), minimization techniques, such as the conjugate gradient method [47], can be used. The function $\vec{r}(u, v)$ and its first-order parametric derivatives can be evaluated as explained in Chapter 2 with adequate values for r and s. Assuming that the surface parametric coordinates vary between 0 and 1, the solutions (u_0 and v_0) outside of this range must be discarded. If this distance ray-surface is zero, the ray intersects the regular surface.

The above method is adequate for infinite-length rays, that is, rays that start at the source and propagate as far as infinity. In the case of finite-length rays (from the source $\vec{S}$ to a final point $\vec{O}$), the method is slightly different. Now, the direction of the ray must be calculated as follows:

$$\hat{v} = \frac{\vec{O} - \vec{S}}{|\vec{O} - \vec{S}|} \tag{3.31}$$

Moreover, an additional test must be performed to determine if the intersection point $\vec{r}(u_0, v_0)$ is between the end points of the ray ($\vec{S}$ and $\vec{O}$) as seen in Figure 3.5.

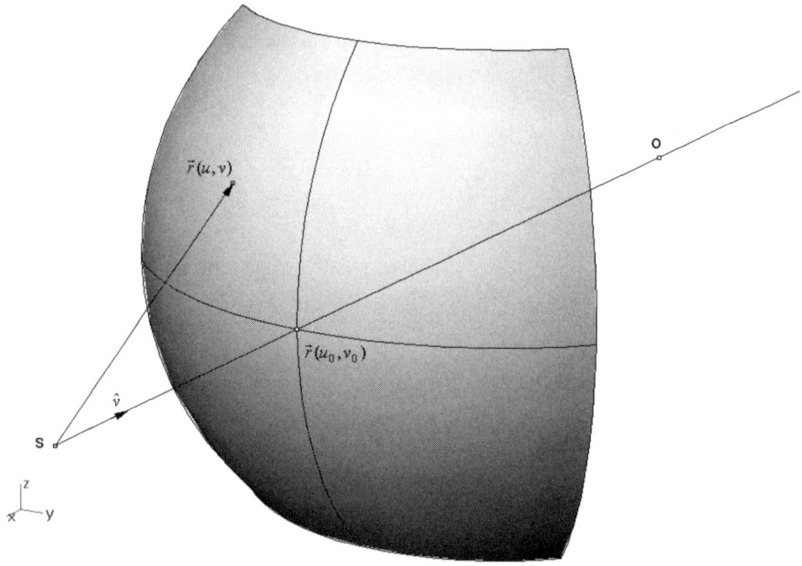

Figure 3.5 Intersection ray-surface.

3.4 REFLECTED FIELD

3.4.1 Electromagnetic Field Calculation

The reflected field at an observation point $\vec{O}$ can be obtained by applying the following expression

$$\vec{E}^r(\vec{O}) = \vec{E}^i(Q_r) \cdot \bar{R} \cdot \sqrt{\frac{\rho_1^r \rho_2^r}{(\rho_1^r + s^r)(\rho_2^r + s^r)}} e^{-jks^r} \tag{3.32}$$

where Q_r is the reflection point, $\vec{E}^i(Q_r)$ is the incident field in the reflection point that is obtained by applying (3.26), s^r is the distance from the reflection point to the observation point, $\bar{R}$ is the reflection coefficient matrix, and ρ_1^r and ρ_2^r are the principal radii of curvature of the reflected wavefront at Q_r.

To apply (3.32), the incident electric field $\vec{E}^i$ at Q_r must be resolved into components that are parallel and perpendicular to the plane of incidence. The plane of incidence is the plane that contains the unit vectors $\hat{n}$ and $\hat{s^i}$, where $\hat{n}$ is the normal vector at the reflection point and $\hat{s^i}$ is the unit vector in the direction of propagation. Similarly, the reflected electric field $\vec{E}^r$ at Q_r can be resolved into components that are parallel and perpendicular to the plane of reflection. The plane of reflection is the plane that contains the unit vectors $\hat{n}$ and $\hat{s^r}$, where $\hat{s^r}$ indicates the direction of the reflected ray at Q_r. Figure 3.6 shows the incidence and reflection planes.

In Figure 3.7, the components parallel and perpendicular to the incident and reflection planes shown in Figure 3.6 can be observed: $\hat{e}^i_{\parallel}$ and $\hat{e}^r_{\parallel}$ are the unit vectors that are parallel to the incident and reflected planes, respectively, and $\hat{e}^i_{\perp}$ and $\hat{e}^r_{\perp}$ are the unit vectors that are perpendicular to the incident and reflected planes, respectively. These vectors are obtained according to the following:

$$\hat{e}^i_{\perp} \times \hat{s^i} = \hat{e}^i_{\parallel} \tag{3.33}$$

$$\hat{e}^r_{\perp} \times \hat{s^r} = \hat{e}^r_{\parallel} \tag{3.34}$$

$$\hat{e}^i_{\parallel} = \frac{\hat{s^i} \times (\hat{n} \times \hat{s^i})}{|\hat{s^i} \times (\hat{n} \times \hat{s^i})|} \tag{3.35}$$

$$\hat{e}^r_{\parallel} = \frac{\hat{s^r} \times (\hat{n} \times \hat{s^r})}{|\hat{s^r} \times (\hat{n} \times \hat{s^r})|} \tag{3.36}$$

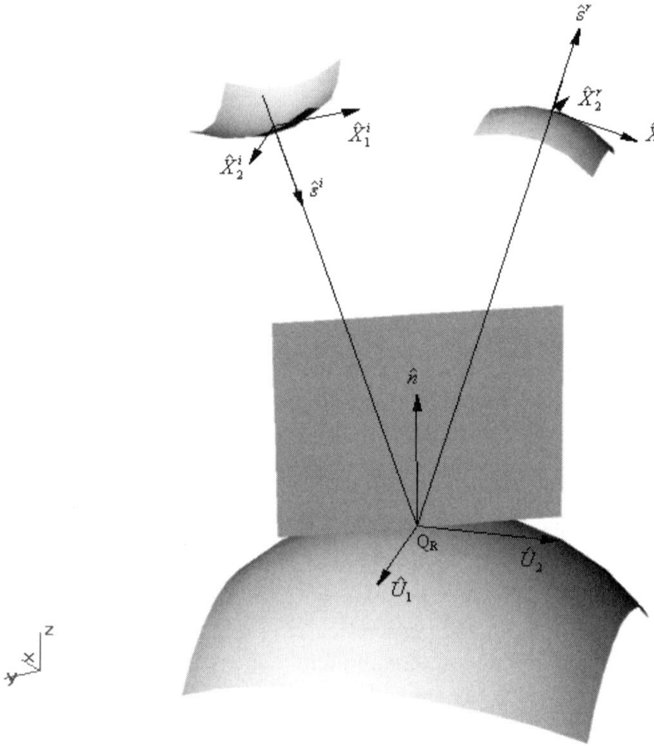

Figure 3.6 Geometric parameters associated with the reflection on a surface.

The angle of incidence θ_i is measured in the plane of incidence and is given by

$$\theta_i = -\cos^{-1}(\hat{n} \cdot \hat{s^i}) \tag{3.37}$$

The angle of reflection θ_r is measured in the reflection plane and is given by

$$\theta_r = -\cos^{-1}(\hat{n} \cdot \hat{s^r}) \tag{3.38}$$

The law of reflection provides information on the relationship among $\hat{n}$, $\hat{s^i}$ and $\hat{s^r}$, and appeals to fundamental electromagnetic theory [3]:

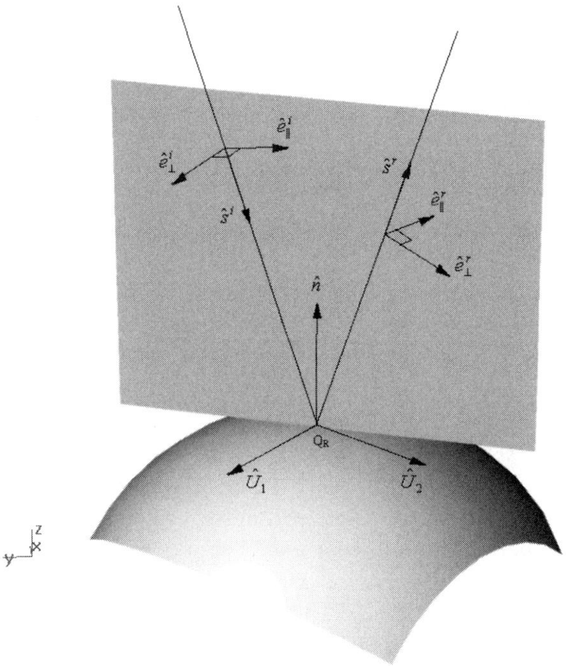

Figure 3.7 Reference system associated with the incident and reflected rays.

$$\hat{n} \cdot \hat{s^i} = -\hat{n} \cdot \hat{s^i} \tag{3.39}$$

Comparing (3.37) with (3.38) and (3.39),

$$\cos(\theta_i) = \cos(\theta_r) \tag{3.40}$$

Thus,

$$\theta_i = \theta_r \tag{3.41}$$

This equation is sometimes referred to as Snell's law of reflection. Therefore, it can be concluded that $\hat{n}$, $\hat{s^i}$ and $\hat{s^r}$ are coplanar. Thus, the plane of incidence and reflection are the same at any point of reflection.

The reflected GO field at the point of reflection is determined by

$$\vec{E}^r(Q_r) = \vec{E}^i(Q_r) \cdot \bar{R} \tag{3.42}$$

Because the fields can be resolved into their parallel and perpendicular components, expression (3.42) can be expressed as a matrix:

$$\begin{bmatrix} \vec{E}_{\parallel}^r \\ \vec{E}_{\perp}^r \end{bmatrix} = \bar{R} \cdot \begin{bmatrix} \vec{E}_{\parallel}^i \\ \vec{E}_{\perp}^i \end{bmatrix} \tag{3.43}$$

where $\vec{E}_{\parallel}^i$ and $\vec{E}_{\perp}^i$ are the components that are parallel and perpendicular, respectively, to the incident field at Q_r, and $\vec{E}_{\parallel}^r$ and $\vec{E}_{\perp}^r$ are the components that are parallel and perpendicular, respectively, to the reflected field at Q_r. The reflection matrix can be written as follows

$$\bar{R} = \begin{bmatrix} R_h & 0 \\ 0 & R_s \end{bmatrix} \tag{3.44}$$

where R_h is the parallel reflection coefficient (known as hard) and R_s is the perpendicular reflection coefficient (known as soft). For a perfectly conductive surface, $R_h = 1$ and $R_s = -1$. Otherwise, the coefficients can be obtained as shown in [40].

The expressions for the principal radii of curvature of the reflected wavefront at Q_r were obtained by Kouyoumjian and Pathak [10]:

$$\frac{1}{\rho_1^r} = \frac{1}{2}\left(\frac{1}{\rho_1^i} + \frac{1}{\rho_2^i}\right) + \frac{1}{f_1} \tag{3.45}$$

$$\frac{1}{\rho_2^r} = \frac{1}{2}\left(\frac{1}{\rho_1^i} + \frac{1}{\rho_2^i}\right) + \frac{1}{f_2} \tag{3.46}$$

where ρ_1^i is the radius of curvature 1 of the incident wavefront at the reflection point (Q_r), ρ_2^i is the radius of curvature 2 of the incident wavefront at the reflection point (Q_r), and f_1 and f_2 are the focal distances.

$$
\begin{aligned}
\frac{1}{f_{1,2}} &= \frac{\cos\theta^i}{|\theta|^2}\left(\frac{\theta_{22}^2+\theta_{12}^2}{a_1}+\frac{\theta_{21}^2+\theta_{11}^2}{a_2}\right) \pm \left\{\left(\frac{1}{\rho_1^i}-\frac{1}{\rho_2^i}\right)^2\right. \\
&+ \left(\frac{1}{\rho_1^i}-\frac{1}{\rho_2^i}\right)\frac{4\cos\theta^i}{|\theta|^2}\left(\frac{\theta_{22}^2+\theta_{12}^2}{a_1}+\frac{\theta_{21}^2+\theta_{11}^2}{a_2}\right) \\
&+ \left.\frac{4\cos\theta^i}{|\theta|^4}\left[\left(\frac{\theta_{22}^2+\theta_{12}^2}{a_1}+\frac{\theta_{21}^2+\theta_{11}^2}{a_2}\right)^2-\frac{4|\theta|^2}{a_1a_2}\right]\right\}^{\frac{1}{2}}
\end{aligned}
\tag{3.47}
$$

where a_1 and a_2 are the radii of curvature of the surface at the reflection point Q_r, θ_{jk} is the component jk of the matrix $[\theta]$, and $|\theta|$ is the determinant of the matrix $[\theta]$.

The matrix $[\theta]$ is given by

$$
[\theta] = \begin{bmatrix} \theta_{11} & \theta_{12} \\ \theta_{21} & \theta_{22} \end{bmatrix} = \begin{bmatrix} \hat{X}_1^i\cdot\hat{U}_1 & \hat{X}_1^i\cdot\hat{U}_2 \\ \hat{X}_2^i\cdot\hat{U}_1 & \hat{X}_2^i\cdot\hat{U}_2 \end{bmatrix}
\tag{3.48}
$$

where $\hat{X}_1^i$ is the principal direction 1 of the incident wavefront, $\hat{X}_2^i$ is the principal direction 2 of the incident wavefront, $\hat{U}_1$ is the principal direction 1 of the surface at the reflection point, and $\hat{U}_2$ is the principal direction 2 of the surface at the reflection point. These vectors can be observed in Figure 3.6.

The principal directions of the reflected wavefront are given by

$$
\hat{X}_1^r = \frac{(Q_{22}^r - 1/\rho_1^r)\hat{\sigma}_1 - Q_{12}^r\hat{\sigma}_2}{\sqrt{(Q_{22}^r - 1/\rho_1^r)^2 + (Q_{22}^r)^2}}
\tag{3.49}
$$

$$
\hat{X}_2^r = -\hat{s}^r \times \hat{X}_1^r
\tag{3.50}
$$

where $\hat{\sigma}_1$ and $\hat{\sigma}_2$ are the intermediate vectors and are given by

$$
\hat{\sigma}_1 = \hat{X}_1^i - 2(\hat{n}\cdot\hat{X}_1^i)\hat{n} \qquad\qquad \hat{\sigma}_2 = \hat{X}_2^i - 2(\hat{n}\cdot\hat{X}_2^i)\hat{n}
\tag{3.51}
$$

The matrix of curvatures Q is defined as the following:

$$
Q = \begin{bmatrix} Q_{11}^r & Q_{12}^r \\ Q_{21}^r & Q_{22}^r \end{bmatrix} = \begin{bmatrix} \dfrac{1}{\rho_1^i} + \dfrac{2}{a_t\cos\theta_i} & -\sin(2\alpha)\left[\dfrac{1}{a_1}-\dfrac{1}{a_2}\right] \\ -\sin(2\alpha)\left[\dfrac{1}{a_1}-\dfrac{1}{a_2}\right] & \dfrac{1}{\rho_1^i} + \dfrac{2\cos\theta_i}{a_t} \end{bmatrix}
\tag{3.52}
$$

When the reflected field is observed in the far zone, we have $s^r \gg \rho_1^r$ and $s^r \gg \rho_2^r$ and the amplitude term of expression (3.23) is

$$\frac{\sqrt{\rho_1^r \rho_2^r}}{s^r} \tag{3.53}$$

3.4.2 Determining Reflected Rays and Computing Reflection Points

Given a NURBS surface, Fermat's principle is used to determine the existence of a reflection point over it. The idea is to minimize the distance of the reflected ray. In far-field analysis, the reflection points, in a given surface, can be calculated by minimizing the following function

$$d(u,v) = d_1(u,v) + d_2(u,v) = |\vec{s}(u,v)| + (D - \hat{v} \cdot \vec{r}(u,v)) \tag{3.54}$$

where the function $d_1(u,v)$ represents the distance from the source $\vec{S}$ to a surface point $\vec{r}(u,v)$, the function $d_2(u,v)$ represents the distance from the surface point to a plane perpendicular to the direction of observation ($\hat{v}$), and D is the independent coefficient of that plane (Figure 3.8). The remaining coefficients of the described plane are the coordinates of $\vec{v}$. The coefficient D can have an arbitrary value, that is, the method works regardless its value. The function $\vec{s}(u,v)$ is defined in (3.30). The partial derivatives of the function $d(u,v)$ are given by

$$\frac{\partial d(u,v)}{\partial u} = \frac{\partial d_1(u,v)}{\partial u} + \frac{\partial d_2(u,v)}{\partial u} = [\hat{s}(u,v) - \hat{v}] \cdot \vec{r}_u(u,v) \tag{3.55}$$

$$\frac{\partial d(u,v)}{\partial v} = \frac{\partial d_1(u,v)}{\partial v} + \frac{\partial d_2(u,v)}{\partial v} = [\hat{s}(u,v) - \hat{v}] \cdot \vec{r}_v(u,v) \tag{3.56}$$

By minimizing $d(u,v)$, the reflection point candidate (u_0, v_0) is obtained. To be a real solution, its parametric values must be in the range [0, 1]. All the terms of (3.54) through (3.56) can be expressed as functions of $\vec{r}(u,v)$ and its first-order parametric derivatives; thus, they can be obtained from the equations presented in Chapter 2.

For near-field analysis, the following function is minimized:

$$d(u,v) = d_1(u,v) + d_2(u,v) = |\vec{s}_1(u,v)| + |\vec{s}_2(u,v)| \tag{3.57}$$

where

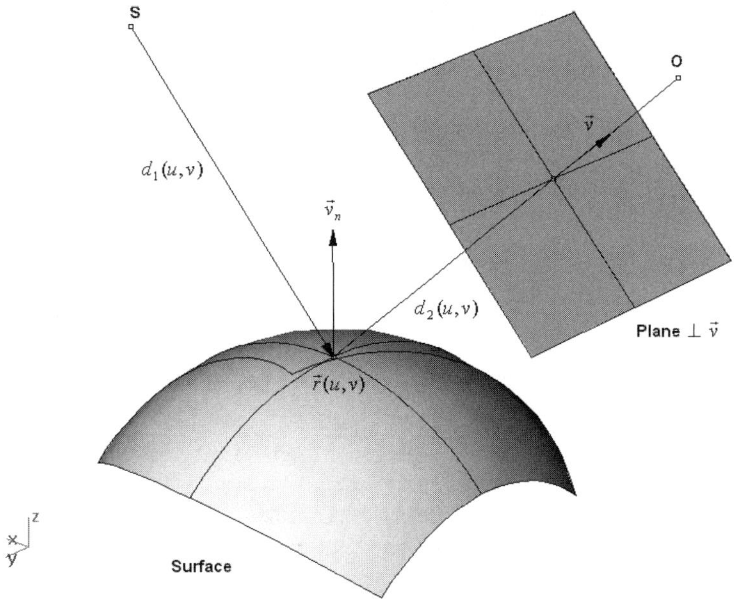

Figure 3.8 Ray-tracing for a reflection.

$$\vec{s}_1(u,v) = \vec{r}(u,v) - \vec{S} \tag{3.58}$$

$$\vec{s}_2(u,v) = \vec{r}(u,v) - \vec{O} \tag{3.59}$$

In (3.57), the first term is the distance from the source $\vec{S}$ to the surface point $\vec{r}(u,v)$, and the second term represents the distance between the surface point and the observation point $\vec{O}$. The partial derivatives of (3.57) are given by

$$\frac{\partial d(u,v)}{\partial u} = [\hat{s}_1(u,v) + \hat{s}_2(u,v)] \cdot \vec{r}_u(u,v) \tag{3.60}$$

$$\frac{\partial d(u,v)}{\partial v} = [\hat{s}_1(u,v) + \hat{s}_2(u,v)] \cdot \vec{r}_v(u,v) \tag{3.61}$$

When a reflection point is obtained, it must be analyzed if the incident and reflected paths of the ray are hidden by any other surface of the model. There are two paths to analyze: the incident path (from $\vec{S}$ to the reflection point $\vec{r}(u_0, v_0)$) and the reflection path. In the far-field case, the reflection path goes from the reflection point $\vec{r}(u_0, v_0)$ to infinity in the direction given by $\hat{v}$. In the near-field case, the reflection path connects the reflection point $\vec{r}(u_0, v_0)$ and the observation point $\vec{O}$. Analyzing the above hidden paths can be accomplished using the ray-surface intersection algorithms of Section 3.3.2.

3.5 DIFFRACTED FIELD

3.5.1 Electromagnetic Field Calculation

The direct and reflected field contributions are predicted by the GO, which is incapable of predicting the field in the shadow zone (see Figure 3.9). To evaluate the field in the shadow zone, we must use the GTD. The uniform formulation of the GTD, that is, the Uniform Theory of Diffraction (UTD) is used to avoid the deficiencies of GTD diffraction coefficients at the shadow boundaries.

According to the GTD, one incident ray causes in an infinite number of diffracted rays that lay in the diffraction cone, or "Keller's cone" [8], with the vertex at the diffraction point. When a ray incident forms an angle β_0' with the edge, the diffracted rays form a cone with half-angle β_0, and $\beta_0' = \beta_0$, as shown in Figure 3.10.

In Figure 3.10, $\hat{n}$ is the unit vector that is normal to the vector of the edge at the diffraction point.

As for GO, the diffracted field at an observation point $\vec{O}$ is determined by applying:

$$\vec{E}^d(\vec{O}) = \vec{E}^i(Q_D) \cdot \bar{D} \cdot \sqrt{\frac{\rho}{s(\rho + s)}} e^{-jks} \tag{3.62}$$

where $\vec{E}^i(Q_D)$ is the incident field at the diffraction point Q_D, which is calculated by applying (3.23), $\bar{D}$ is the diffraction coefficient matrix, s is the distance from the diffraction point Q_D to the observation point, and ρ is the edge caustic distance.

As in the reflection case, the incident field must be resolved into components that are parallel and perpendicular to the edge-fixed plane of incidence. Thus, the diffracted field components are parallel and perpendicular to the edge-fixed plane of diffraction. The edge-fixed coordinate system is shown in Figure 3.11.

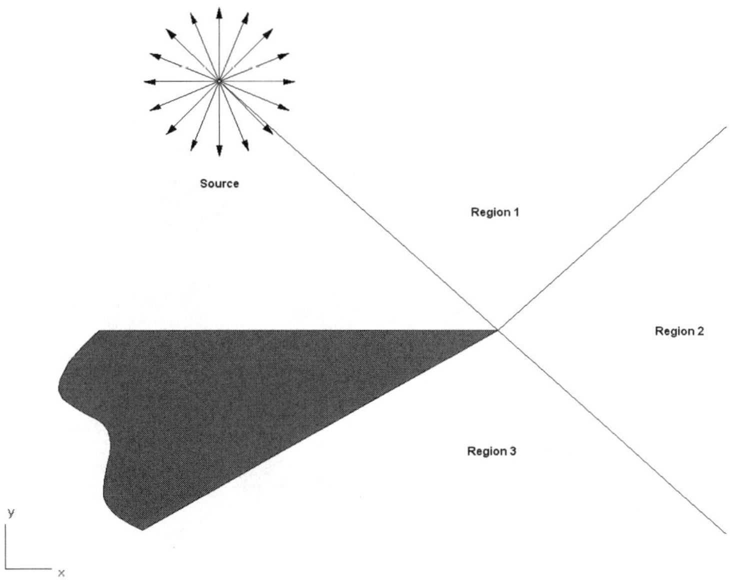

Figure 3.9 Division of the space in regions.

In Figure 3.11, $\hat{e}$ is the unit vector that is tangent to the edge at the diffraction point, $\hat{s}'$ is the unit vector in the incident direction and $\hat{s}$ is the unit vector in the direction of propagation along any one of the diffracted rays. The incident field is resolved into $\hat{\beta}_0'$ and $\hat{\phi}'$ components, and the diffracted field is resolved into $\hat{\beta}_0$ and $\hat{\phi}$ components [44], where

$$\hat{\phi}' = -\frac{\hat{e} \times \hat{s}'}{|\hat{e} \times \hat{s}'|} \tag{3.63}$$

$$\hat{\beta}_0' = \hat{\phi}' \times \hat{s}' \tag{3.64}$$

$$\hat{\phi} = -\frac{\hat{e} \times \hat{s}}{|\hat{e} \times \hat{s}|} \tag{3.65}$$

$$\hat{\beta}_0 = \hat{\phi} \times \hat{s} \tag{3.66}$$

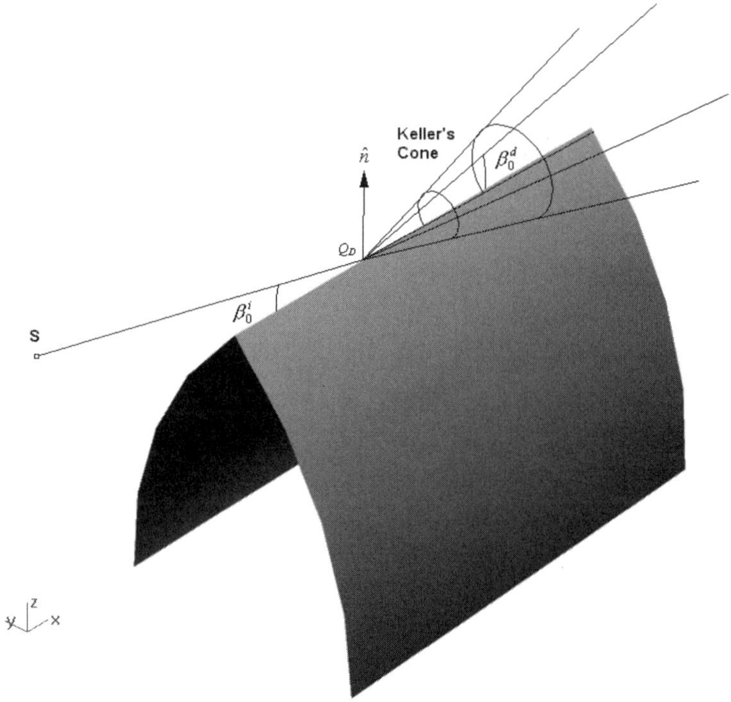

Figure 3.10 Diffraction on an edge; Keller's cone.

The diffracted electric field can be related to the incident electric field at the diffraction point by the diffraction coefficient. The incident and diffracted fields can be expressed in the edge-fixed coordinate system:

$$\begin{bmatrix} \vec{E}^d_{\beta_0} \\ \vec{E}^d_{\phi} \end{bmatrix} = \begin{bmatrix} -D_s & 0 \\ 0 & -D_h \end{bmatrix} \cdot \begin{bmatrix} \vec{E}^i_{\beta_0'} \\ \vec{E}^i_{\phi'} \end{bmatrix} \tag{3.67}$$

The diffraction coefficients are given by:

$$D_{s,h}(L^i, L^{ro}, L^{rn}, \phi, \phi', \beta_0, n) = D_1 + D_2 + R_{s,h}(D_3 + D_4) \tag{3.68}$$

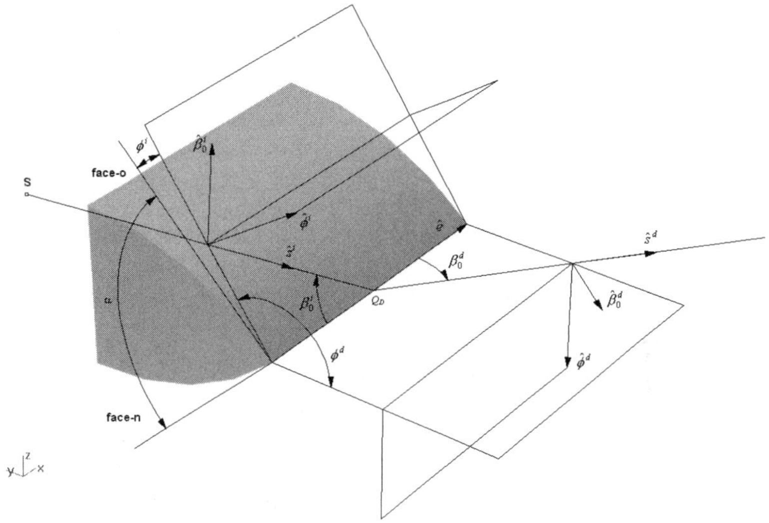

Figure 3.11 Reference system associated with the diffraction.

The components of the diffraction coefficients are given by:

$$D_1 = \frac{-e^{-j\pi/4}}{2n\sqrt{2\pi k}\sin\beta_0}\cot\left[\frac{\pi + (\phi - \phi')}{2n}\right] F[kL^i a^+(\phi - \phi')] \tag{3.69}$$

$$D_2 = \frac{-e^{-j\pi/4}}{2n\sqrt{2\pi k}\sin\beta_0}\cot\left[\frac{\pi - (\phi - \phi')}{2n}\right] F[kL^i a^-(\phi - \phi')] \tag{3.70}$$

$$D_3 = \frac{-e^{-j\pi/4}}{2n\sqrt{2\pi k}\sin\beta_0}\cot\left[\frac{\pi + (\phi + \phi')}{2n}\right] F[kL^{rn} a^+(\phi + \phi')] \tag{3.71}$$

$$D_4 = \frac{-e^{-j\pi/4}}{2n\sqrt{2\pi k}\sin\beta_0}\cot\left[\frac{\pi - (\phi + \phi')}{2n}\right] F[kL^{ro} a^-(\phi + \phi')] \tag{3.72}$$

The L^i are distance parameters that are associated with the incident shadow boundaries and are given, for near field (NF) and far field (FF), by:

$$L^i_{NF} = \frac{s(\rho^i_e + s)\rho^i_1\rho^i_2}{\rho^i_e(\rho^i_1 + s)(\rho^i_2 + s)}\sin^2\beta_0 \qquad L^i_{FF} = \frac{\rho^i_1\rho^i_2}{\rho^i_e}\sin^2\beta_0 \tag{3.73}$$

$$L^{ro,n}_{NF} = \frac{s(\rho^{ro,n}_e + s)\rho^{ro,n}_1\rho^{ro,n}_2}{\rho^{ro,n}_e(\rho^{ro,n}_1 + s)(\rho^{ro,n}_2 + s)}\sin^2\beta_0 \qquad L^i_{FF} = \frac{\rho^{ro,n}_1\rho^{ro,n}_2}{\rho^{ro,n}_e}\sin^2\beta_0 \tag{3.74}$$

where ρ^i_1 is the first radius of curvature of the incident wavefront at the edge; ρ^i_2 is the second radius of curvature of the incident wavefront at the edge; ρ^i_e is the radius of curvature of the incident wavefront at the edge-fixed plane of incidence; $\rho^{ro,n}_1$ is the first radius of curvature of the reflected wavefront from the o and n faces, respectively; $\rho^{ro,n}_2$ is the second radius of curvature of the reflected wavefront from the o and n faces, respectively; and $\rho^{ro,n}_e$ is the radius of curvature of the reflected wavefront from the plane containing the reflected ray and the edge.

The function $a^{\pm}$ of (3.69) through (3.72) is given by

$$a^{\pm}(\beta^{\pm}) = 2\cos^2\left(\frac{2n\pi N^{\pm} - \beta^{\pm}}{2}\right) \tag{3.75}$$

where $\beta^{\pm} = \phi \pm \phi'$, $2n\pi N^{+} - \phi \pm \phi' = \pi$ and $2n\pi N^{-} - \phi \pm \phi' = -\pi$.

The function F of (3.69) through (3.72) represents the value of the Fresnel transition function for any specified argument, which is given by:

$$F[x] = 2j\sqrt{x}e^{jx}\int_{\sqrt{x}}^{\infty} e^{-jr^2}dr \tag{3.76}$$

The edge caustic distance, ρ, is given by

$$\frac{1}{\rho} = \frac{1}{\rho^i_e} - \frac{\hat{n}_e \cdot (\hat{s}' - hats)}{|a_e|\sin^2\beta_0} \tag{3.77}$$

where ρ^i_e is the radius of curvature of the reflected wavefront in the edge-fixed plane of incidence, and a_e is the radius of curvature of the edge at the diffraction point.

When the diffracted field is observed in the far zone, we have $s \gg \rho$, and (3.23) is approximated as [44]

$$\frac{\sqrt{\rho}}{s} \tag{3.78}$$

3.5.2 Determining Diffracted Rays and Computing Diffraction Points

As in the reflection case, Fermat's principle is applied to determine the diffraction point on each edge of the model. The edges correspond to the boundaries of the surfaces in the parametric space ($u = 0$, $u = 1$, $v = 0$, or $v = 1$). Therefore, one of the parametric coordinates of the surface is fixed for a given boundary, being the function to minimize only one variable.

For far-field analysis, the diffraction points on a given edge can be calculated by minimizing the following function:

$$d(t) = d_1(t) + d_2(t) = |\vec{s}(t)| + (D - \hat{v} \cdot \vec{r}(t)) \tag{3.79}$$

where function $d_1(t)$ represents the distance from the source $\vec{S}$ to a point $\vec{r}(t)$ in the edge, function $d_2(t)$ represents the distance from the surface point to a plane that is perpendicular to the direction of observation ($\hat{v}$), and D is the independent coefficient of that plane. The remaining coefficients of such a plane are the coordinates of $\vec{v}$. The coefficient D can have an arbitrary value, but the method works regardless of its value. The function $\vec{s}(t)$ is defined by:

$$\vec{s}(t) = \vec{r}(t) - \vec{S} \tag{3.80}$$

Figure 3.12 illustrates this situation.

The partial derivative of the function $d(t)$ is given by

$$\frac{\partial d(t)}{\partial t} = \frac{\partial d_1(t)}{\partial t} + \frac{\partial d_2(t)}{\partial t} = [\hat{s}(t) - \hat{v}] \cdot \vec{r}_t(t) \tag{3.81}$$

where $\vec{r}_t(t)$ can be calculated as the derivative of the parametric coordinate of the boundary as given in Chapter 2.

Minimizing $d(t)$, the candidate diffraction point (t_0) is obtained. To be a real solution, its parametric values must be in the range of the parametric space of the curve [0,1]. All the terms of (3.79) through (3.81) can be expressed as functions of $\vec{r}(t)$ and its first-order parametric derivative; thus, they can be obtained from the equations given in Chapter 2.

For near-field analysis, the function to minimize is:

$$d(t) = d_1(t) + d_2(t) = |\vec{s}_1(t)| + |\vec{s}_2(t)| \tag{3.82}$$

where

$$\vec{s}_1(t) = \vec{r}(t) - \vec{S} \tag{3.83}$$

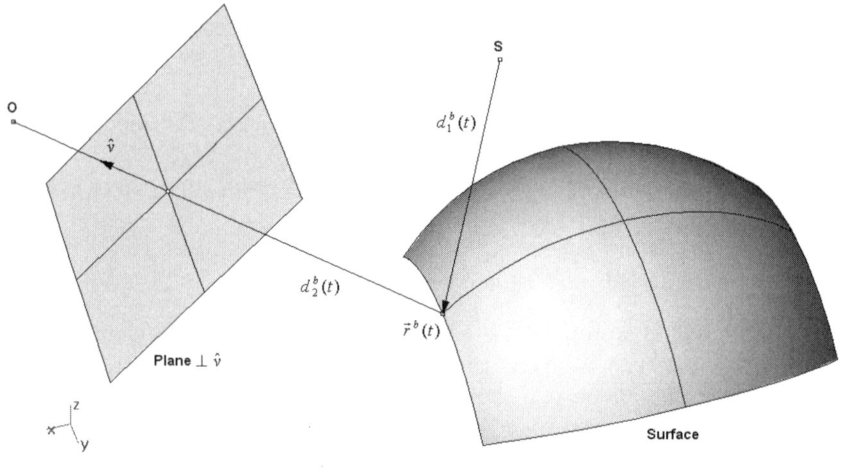

Figure 3.12 Ray-tracing for diffraction.

$$\vec{s}_2(t) = \vec{r}(t) - \vec{O} \tag{3.84}$$

In (3.82), the first term is the distance from the source $\vec{S}$ to a point $\vec{r}(t)$ on the edge, and the second term represents the distance between that point and the observation point $\vec{O}$. The partial derivative of (3.82) is given by:

$$\frac{\partial d(t)}{\partial t} = [\hat{s}_1(t) + \hat{s}_2(t)] \cdot \vec{r}_u(t) \tag{3.85}$$

Once again, the terms of (3.82) through (3.85) are functions of $\vec{r}(t)$ and its first-order parametric derivative; thus, they can be obtained from the equations presented in Chapter 2.

When a diffraction point is obtained, it must be determined whether or not the incident and reflected paths of the ray are hidden by any other surface of the model. There are two paths to analyze: the incident path (from $\vec{S}$ to the diffraction point $\vec{r}(t_0)$) and the diffracted path. In the far-field case, the diffraction path goes from the diffraction point $\vec{r}(t_0)$ to infinity in the direction given by $\hat{v}$. In the near-field case, the diffraction path connects the diffraction point $\vec{r}(t_0)$ and the observation point $\vec{O}$. Analyzing the above hidden paths can be accomplished using the ray-surface intersection algorithms presented in Section 3.3.2.

3.6 DOUBLE EFFECTS

3.6.1 Electromagnetic Field Calculation

Double effects are sequential combinations of two of the effects presented in previous sections: reflection and diffraction. The resulting electric field is calculated by applying the equations of the effects involved sequentially.

For the double reflection case, the first step is to calculate the incident field at the first reflection point by applying (3.26), the second step is to calculate the incident field at the second reflection point by applying (3.32), and finally, the last step is to calculate the electric field at the observation point or in the observation direction by applying (3.32) again.

For the reflection-diffraction case, the first step is to calculate the incident field at the reflection point by applying (3.26), the second step is to calculate the incident field at the diffraction point by applying (3.32), and finally, the last step is to calculate the electric field at the observation point or in the observation direction by applying (3.62).

For the diffraction-reflection case, the first step is to calculate the incident field at the diffraction point by applying (3.26), the second step is to calculate the incident field at the reflection point by applying (3.62), and finally, the last step is to calculate the electric field at the observation point or in the observation direction by applying (3.32).

For the double diffraction case, the first step is to calculate the incident field at the first diffraction point by applying (3.26), the second step is to calculate the incident field at the second diffraction point by applying (3.62), and finally, the last step is to calculate the electric field at the observation point or in the observation direction by applying (3.62) again.

3.6.2 Determining Double Effect Rays

To determine the flash points (reflection or diffraction points), the procedure is very similar to that for simple effects. The distance function must be established, and according to Fermat's principle, this function must be minimized with respect to the variables that appear in the distance function.

3.6.2.1 Double Reflection

For far-field analysis, the reflection points can be calculated by minimizing the following function:

$$\begin{aligned} d(u_1,v_1,u_2,v_2) &= d_1(u_1,v_1)+d_2(u_1,v_1,u_2,v_2)+d_3(u_2,v_2) = \\ &= |\vec{s}_1(u_1,v_1)|+|\vec{s}_2(u_1,v_1,u_2,v_2)|+(D-\hat{v}\cdot\vec{r}_2(u_2,v_2)) \end{aligned} \tag{3.86}$$

where (u_1,v_1) are the parametric coordinates of the first surface, (u_2,v_2) are the parametric coordinates of the second surface, the function $d_1(u_1,v_1)$ represents the distance from the source $\vec{S}$ to a point $\vec{r}_1(u_1,v_1)$ in the first surface, the function $d_2(u_1,v_1,u_2,v_2)$ represents the distance from a point $\vec{r}_1(u_1,v_1)$ in the first surface to a point $\vec{r}_2(u_2,v_2)$ in the second surface, $d_3(u_2,v_2)$ represents the distance from $\vec{r}_2(u_2,v_2)$ to the plane that is perpendicular to the direction of observation ($\hat{v}$), and D is the independent coefficient of that plane. The remaining coefficients of the plane are the coordinates of $\vec{v}$. The coefficient D can have an arbitrary value, that is, the method works regardless its value. The functions $\vec{s}_1(u_1,v_1)$ and $\vec{s}_2(u_1,v_1,u_2,v_2)$ are defined as follows:

$$\vec{s}_1(u_1,v_1)=\vec{r}_1(u_1,v_1)-\vec{S} \tag{3.87}$$

$$\vec{s}_2(u_1,v_1,u_2,v_2)=\vec{r}_2(u_2,v_2)-\vec{r}_1(u_1,v_1) \tag{3.88}$$

For near-field analysis, the function to minimize is the following:

$$\begin{aligned} d(u_1,v_1,u_2,v_2) &= d_1(u_1,v_1)+d_2(u_1,v_1,u_2,v_2)+d_3(u_2,v_2) = \\ &= |\vec{s}_1(u_1,v_1)|+|\vec{s}_2(u_1,v_1,u_2,v_2)|+|\vec{s}_3(u_2,v_2)| \end{aligned} \tag{3.89}$$

where

$$\vec{s}_1(u_1,v_1)=\vec{r}_1(u_1,v_1)-\vec{S} \tag{3.90}$$

$$\vec{s}_2(u_1,v_1,u_2,v_2)=\vec{r}_2(u_2,v_2)-\vec{r}_1(u_1,v_1) \tag{3.91}$$

$$\vec{s}_3(u_2,v_2)=\vec{r}_2(u_2,v_2)-\vec{O} \tag{3.92}$$

In (3.89) the function $d_1(u_1, v_1)$ represents the distance from the source $\vec{S}$ to a point $\vec{r}_1(u_1, v_1)$ in the first surface, the function $d_2(u_1, v_1, u_2, v_2)$ represents the distance from a point $\vec{r}_1(u_1, v_1)$ in the first surface to a point $\vec{r}_2(u_2, v_2)$ in the second surface, and $d_3(u_2, v_2)$ represents the distance from $\vec{r}_2(u_2, v_2)$ in the second surface to the observation point $\vec{O}$.

3.6.2.2 Reflection-Diffraction

For far-field analysis, the reflection and diffraction points can be calculated by minimizing the following function:

$$d(u, v, t) = d_1(u, v) + d_2(u, v, t) + d_3(t) = |\vec{s}_1(u, v)| + |\vec{s}_2(u, v, t)| + (D - \hat{v} \cdot \vec{r}_2(t)) \tag{3.93}$$

where (u, v) are the parametric coordinates of the surface, t is the parametric coordinate of the edge, the function $d_1(u, v)$ represents the distance from the source $\vec{S}$ to a point $\vec{r}_1(u, v)$ in the surface, the function $d_2(u, v, t)$ represents the distance from a point $\vec{r}_1(u, v)$ in the surface to a point $\vec{r}_2(t)$ in the edge, $d_3(t)$ represents the distance from $\vec{r}_2(t)$ to a plane perpendicular to the direction of observation ($\hat{v}$), and D is the independent coefficient of that plane. The remaining coefficients of that plane are the coordinates of $\vec{v}$. The coefficient D can have an arbitrary value, that is, the method works regardless its value. The functions $\vec{s}_1(u, v)$ and $\vec{s}_2(u, v, t)$ are defined as follows:

$$\vec{s}_1(u, v) = \vec{r}_1(u, v) - \vec{S} \tag{3.94}$$

$$\vec{s}_2(u, v, t) = \vec{r}_2(t) - \vec{r}_1(u, v) \tag{3.95}$$

For near-field analysis, the function to minimize is the following:

$$d(u, v, t) = d_1(u, v) + d_2(u, v, t) + d_3(t) = |\vec{s}_1(u, v)| + |\vec{s}_2(u, v, t)| + |\vec{s}_3(t)| \tag{3.96}$$

where

$$\vec{s}_1(u, v) = \vec{r}_1(u, v) - \vec{S} \tag{3.97}$$

$$\vec{s}_2(u, v, t) = \vec{r}_2(t) - \vec{r}_1(u, v) \tag{3.98}$$

$$\vec{s}_3(t) = \vec{r}_2(t) - \vec{O} \tag{3.99}$$

In (3.96), the function $d_1(u,v)$ represents the distance from the source $\vec{S}$ to a point $\vec{r_1}(u,v)$ in the surface, the function $d_2(u,v,t)$ represents the distance from a point $\vec{r_1}(u,v)$ in the surface to a point $\vec{r_2}(t)$ in the edge, and $d_3(t)$ represents the distance from $\vec{r_2}(t)$ to the observation point $\vec{O}$.

3.6.2.3 Diffraction-Reflection

For far-field analysis, the diffraction and reflection points can be calculated by minimizing the following function:

$$d(t,u,v) = d_1(t)+d_2(t,u,v)+d_3(u,v) = |\vec{s_1}(t)|+|\vec{s_2}(t,u,v)|+(D-\hat{v}\cdot\vec{r_2}(u,v)) \tag{3.100}$$

where t is the parametric coordinate of the edge, (u,v) are the parametric coordinates of the surface, $d_1(t)$ represents the distance from the source $\vec{S}$ to a point $\vec{r_1}(t)$ in the edge, $d_2(t,u,v)$ represents the distance from a point $\vec{r_1}(t)$ in the edge to a point $\vec{r_2}(u,v)$ in the surface, $d_3(u,v)$ represents the distance from $\vec{r_2}(u,v)$ to a plane perpendicular to the direction of observation ($\hat{v}$), and D is the independent coefficient of that plane. The remaining coefficients of that plane are the coordinates of $\vec{v}$. The coefficient D can have an arbitrary value, that is, the method works regardless its value. The functions $\vec{s_1}(t)$ and $\vec{s_2}(t,u,v)$ are defined as follows:

$$\vec{s_1}(t) = \vec{r_1}(t) - \vec{S} \tag{3.101}$$

$$\vec{s_2}(t,u,v) = \vec{r_2}(u,v) - \vec{r_1}(t) \tag{3.102}$$

For near-field analysis, the function to minimize is the following:

$$d(t,u,v) = d_1(t) + d_2(t,u,v) + d_3(u,v) = |\vec{s_1}(t)| + |\vec{s_2}(t,u,v)| + |\vec{s_3}(u,v)| \tag{3.103}$$

where

$$\vec{s_1}(t) = \vec{r_1}(t) - \vec{S} \tag{3.104}$$

$$\vec{s_2}(t,u,v) = \vec{r_2}(u,v) - \vec{r_1}(t) \tag{3.105}$$

$$\vec{s_3}(u,v) = \vec{r_2}(u,v) - \vec{O} \tag{3.106}$$

In (3.102), the function $d_1(t)$ represents the distance from the source $\vec{S}$ to a point $\vec{r_1}(t)$ in the edge, $d_2(t,u,v)$ represents the distance from a point $\vec{r_1}(t)$ in the edge to a point $\vec{r}_2(u, v)$ in the surface, and $d_3(u, v)$ represents the distance from $\vec{r}_2(u, v)$ to the observation point $\vec{O}$.

3.6.2.4 Double Diffraction

For far-field analysis, the diffraction and reflection points can be calculated by minimizing the following function:

$$d(t_1, t_2) = d_1(t_1) + d_2(t_1, t_2) + d_3(t_2) = |\vec{s}_1(t_1)| + |\vec{s}_2(t_1, t_2)| + (D - \hat{v} \cdot \vec{r}_2(t_2) \tag{3.107}$$

where t_1 is the parametric coordinate of the first edge, t_2 is the parametric coordinate of the second edge, $d_1(t_1)$ represents the distance from the source $\vec{S}$ to a point $\vec{r}_1(t_1)$ in the first edge, $d_2(t_1, t_2)$ represents the distance from a point $\vec{r}_1(t_1)$ in the first edge to a point $\vec{r}_2(t_2)$ in the second edge, $d_3(t_2)$ represents the distance from $\vec{r}_2(t_2)$ to a plane perpendicular to the direction of observation ($\hat{v}$), and D is the independent coefficient of that plane. The remaining coefficients of that plane are the coordinates of $\vec{v}$. The coefficient D can have an arbitrary value, that is, the method works regardless its value. The functions $\vec{s}_1(t_1)$ and $\vec{s}_2(t_1, t_2)$ are defined as follows:

$$\vec{s}_1(t_1) = \vec{r}_1(t_1) - \vec{S} \tag{3.108}$$

$$\vec{s}_2(t_1, t_2) = \vec{r}_2(t_1, t_2) - \vec{r}_1(t_1) \tag{3.109}$$

For near-field analysis, the function to minimize is the following:

$$d(t_1, t_2) = d_1(t_1) + d_2(t_1, t_2) + d_3(t_2) = |\vec{s}_1(t_1)| + |\vec{s}_2(t_1, t_2)| + |\vec{s}_3(t_2)| \tag{3.110}$$

where

$$\vec{s}_1(t_1) = \vec{r}_1(t_1) - \vec{S} \tag{3.111}$$

$$\vec{s}_2(t_1, t_2) = \vec{r}_2(t_1, t_2) - \vec{r}_1(t_1) \tag{3.112}$$

$$\vec{s}_3(t_2) = \vec{r}_2(t_2) - \vec{O} \tag{3.113}$$

In (3.110) the function $d_1(t_1)$ represents the distance from the source $\vec{S}$ to a point $\vec{r}_1(t_1)$ in the first edge, $d_2(t_1, t_2)$ represents the distance from a point $\vec{r}_1(t_1)$ in the first edge to a point $\vec{r}_2(t_2)$ in the second edge, and $d_3(t_2)$ represents the distance from $\vec{r}_2(t_2)$ to the observation point $\vec{O}$.

In all cases, the partial derivatives can be computed as explained in Chapter 2 and in the sections dedicated to simple reflection and diffraction of this chapter. Moreover, in all cases, the possibility of any hidden ray paths must be checked following the procedure described in Section 3.3.2.

3.7 CREEPING WAVES

3.7.1 Electromagnetic Field Calculation

The creeping wave or surface diffracted ray is a phenomenon that appears when a ray illuminates a curved surface at grazing incidence, whose propagating direction is tangent to the curved surface and propagates along the surface following geodesic paths from the incident point. Figure 3.13 shows an example of a creeping wave trajectory. The starting point of the geodesic path is placed at the boundary between the illuminated and the shadowed parts of the surface, taking the source (S in the figure) as the reference, and the end point of this trajectory is also at the boundary between the illuminated and the shadowed parts of the surface, but this time, taking the observer (O in the figure) as the reference.

To evaluate the field produced by the creeping wave phenomenon, it is necessary to know the field at the point Q_2; given the field at that point, the diffracted field is calculated by applying the amplitude and phase terms of the GTD as in edge diffraction, that is,

$$\vec{E}^d(\vec{O}) = \vec{E}^d(Q_2)\frac{e^{-jks^d}}{s^d} \tag{3.114}$$

where $\vec{E}^d(\vec{O})$ is the field produced by the creeping wave phenomenon at the point $\vec{O}$, $\vec{E}^d(Q_2)$ is the diffracted field at the point Q_2, and s^d is the distance between Q_2 and the observer.

The diffracted field at the point Q_2 has the following expression [44]:

$$\vec{E}^d(Q_2) = \vec{E}^i(Q_1)\bar{\bar{T}} \tag{3.115}$$

where $\vec{E}^i(Q_1)$ is the incident field at the point Q_1, and $\bar{\bar{T}}$ is the surface diffraction coefficient matrix.

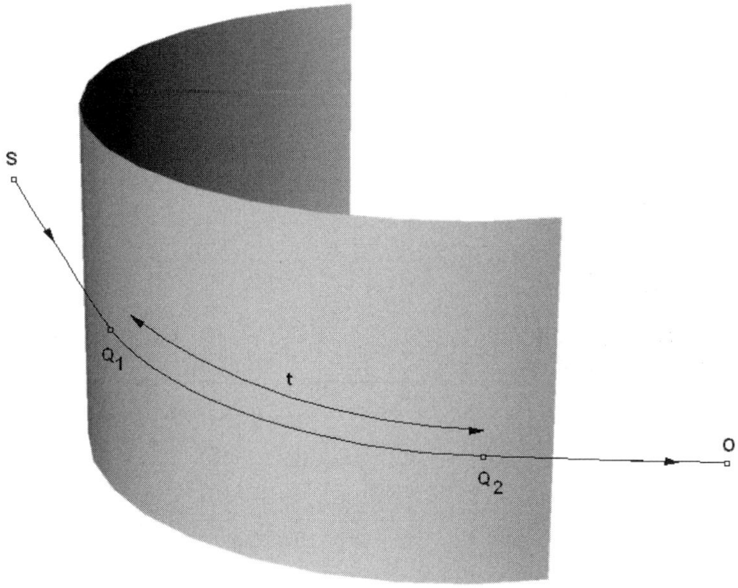

Figure 3.13 Creeping wave phenomenon.

For the radiated field computation, it is necessary to define polarization vectors over the points Q_1 and Q_2. These vectors are obtained from the following expressions:

$$\hat{b}_1 = \hat{t}_1 \times \hat{n}_1 \tag{3.116}$$

$$\hat{b}_2 = \hat{t}_2 \times \hat{n}_2 \tag{3.117}$$

where $\hat{n}_1$, $\hat{n}_2$, $\hat{t}_1$, and $\hat{t}_2$ are the normal and tangential vectors to the surfaces at points Q_1 and Q_2. These vectors are shown in Figure 3.14.

Once the points over the surface and known polarization vectors are fixed, the convex surface diffraction coefficients are calculated. Their expression is:

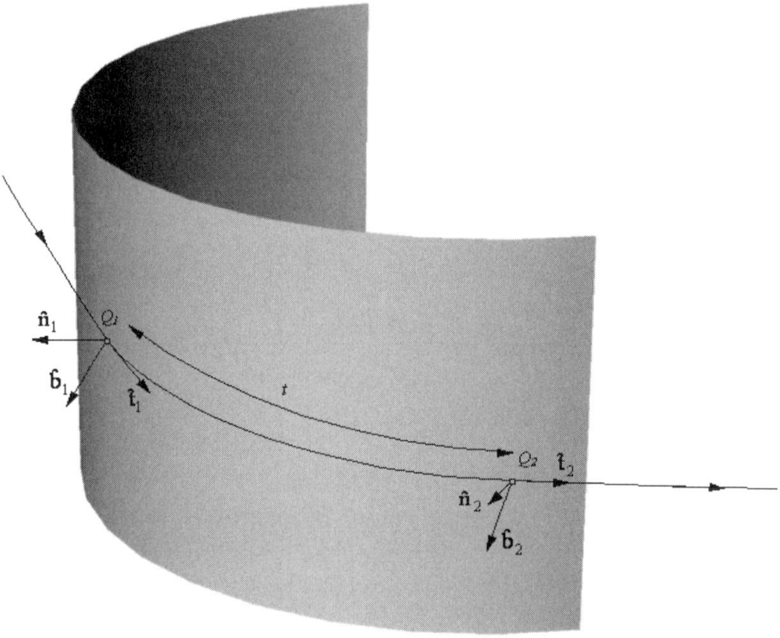

Figure 3.14 Polarization vector for the creeping waves.

$$\begin{aligned} T_{s,h} &= -\left[\sqrt{m(Q_1)m(Q_2)}\sqrt{\frac{2}{k}}\left(\frac{e^{-j(\pi/4}}{2\sqrt{\pi\zeta^d}}\left[1 - F(X^d)+\right]\right.\right. \\ &+ \left.\left.\hat{P}_{s,h}(\zeta^d)\right)\right]\sqrt{\frac{s^i}{s^i+t}}e^{-jkt} \end{aligned} \tag{3.118}$$

where $m(Q_i)$ is a parameter that depends on the surface curvature at the point Q_i as follows:

$$m(Q_i) = \left(\frac{k\rho_g(Q_i)}{2}\right)^{1/3} \tag{3.119}$$

where $\rho_g(Q_i)$ is the radius of curvature at the point Q_i, s^i is the distance between the source and Q_1, and $F(X^d)$ is the Fresnel transition function. The argument X^d is calculated as follows:

$$X^d = \frac{kL^d(\zeta^d)^2}{2m(Q_1)m(Q_2)} \tag{3.120}$$

where L^d is the distance parameter that is calculated as follows:

$$L^d = \frac{s^d s^i}{s^d + s^i} \tag{3.121}$$

The Fock parameter is ζ^d, which is the following in this case:

$$\zeta^d = \int_{Q_1}^{Q_2} dt' \frac{m(t')}{\rho_g(t')} \tag{3.122}$$

where $t = \displaystyle\int_{Q_1}^{Q_2} dt'$ is the arc length of the ray over the surface, which is obtained as the product between the number of points in the ray trajectory and the distance between them ($\Delta\sigma$).

The previous integral (3.122) is evaluated as the product of the sum of the value of $\dfrac{m(t')}{\rho_g(t')}$ at each point of the trajectory with $\Delta\sigma$.

The function $\hat{P}_{s,h}$ is the Pekeris function, and it can be defined from the Fock functions $p*$ and $q*$ [44] as follows [48]:

$$\hat{P}_{s,h}(\zeta^d) = \left\{ \begin{array}{c} p*(\zeta^d) \\ q*(\zeta^d) \end{array} \right\} e^{-j(\pi/4)} - \frac{e^{-j(\pi/4)}}{2\sqrt{\pi}\zeta^d} \tag{3.123}$$

3.7.2 Determining Creeping Wave Trajectories

The creeping rays must also satisfy Fermat's principle. Therefore, the ray arrives at point Q_1 under grazing incidence, travels on the body surface along a geodesic curve, and leaves the surface (at Q_2) tangentially.

To calculate this trajectory over a NURBS surface, the first step is to obtain a set of sampling points on the shadow boundary (SSB) seen from the source. These points satisfy

$$(\vec{r}(u,v) - \vec{S}) \cdot \vec{n}(u,v) = 0 \tag{3.124}$$

where $\vec{r}(u,v)$ is the position vector of the point, $\vec{S}$ is the position vector of the source, and $\vec{n}(u,v)$ is the normal vector to the surface at the point. Obviously, only points from the partially illuminated surfaces fulfill (3.124). In each of those surfaces, there are infinite solutions that form part of an SSB curve.

The SSB points are obtained as follows. To determine an SSB point in an isoparametric curve $v = constant$, the following function is minimized:

$$f(u) = (\hat{s}(u) \cdot \hat{n}(u))^2 \tag{3.125}$$

where

$$\hat{s}(u) = \hat{r}(u) - \vec{S} \tag{3.126}$$

The partial derivative of the function $f(u)$ is

$$\frac{\partial f(u)}{\partial u} = 2\sqrt{f(u)}(\hat{n}(u) \cdot \hat{s}_1(u) + \hat{s}(u) \cdot \hat{s}_2(u)) \tag{3.127}$$

where

$$\hat{s}_1(u) = \frac{\vec{r}_u(u)}{|\vec{s}(u)|} - \frac{\vec{s}(u)(\vec{s}(u) \cdot \vec{r}_u(u)}{|\vec{s}(u)|^3} \tag{3.128}$$

$$\hat{s}_2(u) = \frac{(\vec{r}_u(u) \times \vec{r}_{uv}(u)) + (\vec{r}_{uu}(u) \times \vec{r}_v(u))}{|\vec{r}_u(u) \times \vec{r}_v(u)|} - \tag{3.129}$$
$$\frac{(\vec{r}_u(u) \times \vec{r}_v(u))\left[(\vec{r}_u(u) \times \vec{r}_v(u)) \cdot ((\vec{r}_u(u) \times \vec{r}_{uv}(u)) + (\vec{r}_{uu}(u) \times \vec{r}_v(u)))\right]}{|\vec{r}_u(u) \times \vec{r}_v(u)|^3}$$

To determine an SSB point in an isoparametric curve $v = constant$, we minimize the following function using the CGM for one variable:

$$f(v) = (\hat{s}(v) \cdot \hat{n}(v))^2 \tag{3.130}$$

where

$$\hat{s}(v) = \hat{r}(v) - \vec{S} \tag{3.131}$$

The partial derivative of the function $f(v)$ is

$$\frac{\partial f(v)}{\partial v} = 2\sqrt{f(v)}(\hat{n}(v) \cdot \hat{s}_1(v) + \hat{s}(v) \cdot \hat{s}_2(v)) \tag{3.132}$$

where

$$\hat{s}_1(v) = \frac{\vec{r}_v(v)}{|\vec{s}(v)|} - \frac{\vec{s}(v)(\vec{s}(v) \cdot \vec{r}_v(v)}{|\vec{s}(v)|^3} \tag{3.133}$$

$$\hat{s}_2(v) = \frac{(\vec{r}_u(v) \times \vec{r}_{vv}(v)) + (\vec{r}_{uv}(v) \times \vec{r}_v(v))}{|\vec{r}_u(v) \times \vec{r}_v(v)|} - \tag{3.134}$$
$$\frac{(\vec{r}_u(v) \times \vec{r}_v(v)) \left[(\vec{r}_u(v) \times \vec{r}_v(v)) \cdot ((\vec{r}_u(v) \times \vec{r}_{vv}(v)) + (\vec{r}_{uv}(v) \times \vec{r}_v(v)))\right]}{|\vec{r}_u(v) \times \vec{r}_v(v)|^3}$$

For each of these SSB points, the corresponding geodesic curves that start at the point and satisfy the following equations [49] are calculated:

$$\frac{d^2u}{d^2\sigma} + \Gamma^1_{11}\left(\frac{du}{d\sigma}\right)^2 + 2\Gamma^1_{12}\frac{du}{d\sigma}\frac{dv}{d\sigma} + \Gamma^1_{22}\left(\frac{dv}{d\sigma}\right)^2 = 0 \tag{3.135}$$

$$\frac{d^2v}{d^2\sigma} + \Gamma^2_{11}\left(\frac{du}{d\sigma}\right)^2 + 2\Gamma^2_{12}\frac{du}{d\sigma}\frac{dv}{d\sigma} + \Gamma^2_{22}\left(\frac{dv}{d\sigma}\right)^2 = 0 \tag{3.136}$$

where Γ^k_{ij} are the Christoffel parameters [49] and σ is the arc length along the curve. The Christoffel parameters are functions of the parametric derivatives of the surfaces; thus, at each point (u, v), they can be calculated easily from the surface description.

Therefore, the following equations are obtained by solving (3.135) and (3.136). These equations allow the parametric coordinates of a set of sample points of the geodesic curve, where each point is separated by a distance $\Delta\sigma$ from the previous point over the trajectory, to be obtained iteratively.

$$\begin{aligned} u_{k+2} &= u_k + 2(u_{k+1} - u_k) - \Gamma^1_{11}(u_{k+1} - u_k)^2 \\ &- 2\Gamma^1_{12}(u_{k+1} - u_k)(v_{k+1} - v_k) - \Gamma^1_{22}(v_{k+1} - v_k)^2 \end{aligned} \tag{3.137}$$

$$\begin{aligned} v_{k+2} &= v_k + 2(v_{k+1} - v_k) - \Gamma^2_{11}(u_{k+1} - u_k)^2 \\ &- 2\Gamma^2_{12}(u_{k+1} - u_k)(v_{k+1} - v_k) - \Gamma^2_{22}(v_{k+1} - v_k)^2 \end{aligned} \tag{3.138}$$

As (3.137) and (3.138) state, to calculate one point in the trajectory, two previous points are needed; therefore, two initial points are needed to apply the algorithm. The first starting point (with position vector $\vec{r}_1$), is, obviously, the outline point for which the propagated ray is being calculated. The second starting point (with position vector $\vec{r}_2$) is obtained by considering that the incidence to the surface has to be tangential to the propagation trajectory for the surface diffraction to exist. Therefore, if the step $\Delta\sigma$ is small enough, the second point can be calculated using the following equation:

$$\vec{r}_2 = \vec{r}_1 + \Delta\sigma \frac{(\vec{r}_1 - \vec{S})}{|\vec{r}_1 - \vec{S}|} \tag{3.139}$$

where $\Delta\sigma$ is chosen taking into account the size of the body, and then method can be applied.

Because of the complexity of the bodies, they are modeled as the union of several surfaces. It is important to calculate the path followed by a ray that crosses from one surface to another efficiently. This occurs when the point obtained by the method is outside the limits of a surface (surface 1) that is joined with another surface (surface 2).

First, ray propagation occurs when the surfaces are joined with continuity. To evaluate this condition, the normal vectors to both surfaces are calculated at the boundary between them. There is continuity between the surfaces if the following criterion is satisfied:

$$\hat{n}_1 \cdot \hat{n}_2 = 1 \tag{3.140}$$

where $\hat{n}_1$ and $\hat{n}_2$ are the normal vectors for each surface.

If the above problem is not satisfied, the creeping wave does not propagate. Otherwise, the following steps are performed:

1. The Cartesian coordinates of the point obtained from the first surface are calculated as if the point was inside the surface.

2. From these Cartesian coordinates, the parametric coordinates over the second surface are calculated.

3. Ray propagation continues in the second surface.

In the second surface, two starting points are needed, and the second is the last point of the ray obtained in the first surface. It is evident that this algorithm provides a solution that is slightly out of the parametric space of the second surface, but it is adequate when the surfaces are joined with continuity and the iteration step is small enough.

The creeping ray leaves the surface when it reaches the shadow boundary that is seen from the observation point (point Q_2). To calculate when the shadow boundary has been reached, two conditions have to be evaluated at the point obtained in each iteration:

1. The receiver is visible from the point, that is,

$$\hat{n}_i \cdot \vec{V}_i \geq 0 \tag{3.141}$$

   where i is the index of the point, $\hat{n}_i$ is the normal vector to the surface in this point, and $\vec{V}_i$ is the output vector that joins this point and the observation point.

2. The output ray is emergent, that is to say, the output vector has the same direction as that of the geodesic curve:

$$(\vec{r}_i - \vec{r}_{i-1}) \cdot \vec{V}_i \geq 0 \tag{3.142}$$

   where $\vec{r}_i$ is the point considered and $\vec{r}_{i-1}$ is the point obtained in the previous iteration.

The previous process is followed for each SSB sampling point. Many trajectories of SSB sampling points result, except for those whose trajectory finds a discontinuity between surfaces. However, not all of them produce a field generated by creeping waves. The field only exists if the output ray is tangential to the geodesic curve or, in other words, if the output vector from the last point of the curve is tangential to this curve. Therefore, only the trajectories that meet the tangential condition for the output ray are valid because they are the only ones that satisfy Fermat's principle.

It is clear that the trajectories whose output direction is exactly tangential (exact solutions) to the curve are not obtained because the method works with a discrete sampling of the SSB curve. From all the trajectories calculated, those with a local minimum in the tangential condition are chosen as approximate solutions. The chosen solutions are exact enough if a fine sampling of points in the SSB is used.

References

[1] R. F. Harrington. *Time-Harmonic Electromagnetic Fields*. McGraw-Hill, 1961.

[2] J. B. Keller. "One hundred years of diffraction theory." *IEEE Transactions on Antennas and Propagation*, AP-33(2):123–136, February 1985.

[3] M. Born and E. Wolf. *Principles of Optics*. Pergamon, 1975.

[4] A. Sommerfield and J. Runge. "Awendung der vektorredchung aufdie grundlangen der geometrischen optik." *Annals Physics*, 35:277–298, 1911.

[5] H. Poincaré. "Sur les intégrales irréguliéres des équations linéaires." *Acta Mathematica*, 8(295-344), 1886.

[6] R. M. Luneberg. *Mathematical Theory of Optics*. Brown University Press, 1944.

[7] M. Kline. "An asymptotic solution of Maxwell's equations." *Communications on Pure Applied Mathematics*, 4:225–262, 1951.

[8] J. B. Keller. "Geometrical theory of diffraction." *J. Opt. Soc. Amer.*, 52:116–130, 1962.

[9] J. B. Keller. "Diffraction by a convex cylinder." *IRE Transactions on Antennas and Propagation*, AP-24:312–321, 1956.

[10] R. C. Kouyoumjian and P. H. Pathak. "A geometrical theory of diffraction for an edge in a perfectly conducting surface." *Proceedings of IEEE*, 62(11):1448–1461, November 1974.

[11] P. C. Clemmow. "Some extensions to the method of integration by steepest descents." *Quart. J. Mech. Appl. Math.*, 3:241–256, 1950.

[12] S. W. Lee and G. A. Deschamps. "A uniform asymptotic theory of electromagnetic diffraction by a curved wedge." *IEEE Transactions on Antennas and Propagation*, AP-24:25–34, January 1976.

[13] F. Oberhettinger. "On asymptotic series for functions occurring in the theory of diffraction of waves by wedges." *J. Math. Phys.*, 34:245–255, 1956.

[14] W. D. Burnside, R. J. Marhefka, and C. L. Yu. "Roll-plane analysis of aircraft antennas." *IEEE Transactions on Antennas and Propagation*, AP-21(6):780–786, November 1973.

[15] W. D. Burnside, M. C. Gilreath, R. J. Marhefka, and C. L. Yu. "A study of KC-135 aircraft antenna patterns." *IEEE Transactions on Antennas and Propagation*, AP-23:309–316, May 1975.

[16] C. L. Yu, W. D. Burnside, and M. C. Gilreath. "Volumetric pattern analysis of airborne antennas." *IEEE Transactions on Antennas and Propagation*, AP-26:636–641, September 1978.

[17] W. D. Burnside, N. Wang, and E. L. Pelton. "Near-field pattern analysis of airborne antennas." *IEEE Transactions on Antennas and Propagation*, AP-28:333–342, May 1980.

[18] C. A. Balanis and Y. B. Cheng. "Antenna radiation modeling for microwave landing system." *IEEE Transactions on Antennas and Propagation*, pp. 490–497, July 1976.

[19] J. J. Kim and W. D. Burnside. "Simulation and analysis of antennas radiating in a complex environment." *IEEE Transactions on Antennas and Propagation*, AP-34(4):554–562, April 1986.

[20] J. Pérez, J. A. Saiz, O. M. Conde, R. P. Torres, and M. F. Cátedra. "Analysis of antennas on board arbitrary structures modeled by NURBS surfaces." *IEEE Transactions on Antennas and Propagation*, AP-45(6):1045–1053, June 1997.

[21] J. Pérez and M. F. Cátedra. "Application of physical optics to the RCS computation of bodies modeled with NURBS surfaces." *IEEE Transactions on Antennas and Propagation*, AP-42(10):1404–1411, October 1994.

[22] F. Jensen, S. Mourgues, and M. Sabbadini. "GTD ray tracing by mapping." *AP-S International Symposium*, pp. 21–26, 1996.

[23] K. Chamberlin. "An automated approach for implementing GTD to model 2-D terrain effects at microwave frequencies." *IEEE Transactions on Electromagnetic Compatibility*, 38(1):7–14, February 1996.

[24] M. F. Cátedra, I. González, and L. Lozano. "Fast ray-tracing for computing n-bounces between curved surfaces." *AP-S International Symposium*, 2007.

[25] F. Saez de Adana, O. Gutiérrez, I. González, J. Pérez, and M. F. Cátedra. "General method for the ray tracing on convex bodies." *Applied Computational Electromagnetics Society Journal*, 16(1):20–26, March 2001.

[26] Y. C. Ruan, X. Y. Zhou, J. Y. Chin, T. J. Cui, Y. B. Tao, and H. Lin. "The UTD analysis to EM scattering by arbitrarily convex objects using ray tracing of creeping waves on numerical meshes." *AP-S International Symposium*, 2008.

[27] L. de Haro and J. L. Besada. "Aperture radiation analysis including plane finiteness and mutual coupling." *AP-S International Symposium*, pp. 1572–1575, 1989.

[28] R. Lenormand, C. Rigal, E. Melet, and Y. Beniguel. "Characterisation of the radiation from low frequency antennas on board satellites." *JINA 94*, pp. 66–70, 1994.

[29] M. Anstey and S. A. Saoudy. "Radiation characteristics of a ship-mounted high frequency ground wave radar antenna." *IEEE Antennas and Propagation Society International Symposium*, pp. 1852–1855, 1996.

[30] S. Kashyap and A. Louie. "Electromagnetic analysis of antennas on a ship." *Symposium on Antenna Technology and Applied Electromagnetics*, pp. 771–776, 1996.

[31] S. A. Saoudy and S. K. Lee. "Correlating numerical and scale modelled experimental radiation characteristics of a ship-mounted high frequency antenna." *Symposium on Antenna Technology and Applied Electromagnetics 1996 Conference*, pp. 121–122, 1996.

[32] K. Allsebrook and J. D. Parsons. "Mobile radio propagation in British cities at frequencies in the VHF and UHF bands." *IEEE Proceedings*, 124(2):95–102, 1977.

[33] F. Ikegami, S. Yoshoida, T. Takeuchi, and M. Umehira. "Propagation factors controlling mean field strength on urban streets." *IEEE Transactions on Antennas and Propagation*, AP-32:822–829, December 1984.

[34] J. Walfisch and H. L. Bertoni. "A theoretical model of UHF propagation in urban environments." *IEEE Transactions on Antennas and Propagation*, AP-36:1788–1796, October 1988.

[35] M. F. Cátedra, J. Pérez, F. Saez de Adana, and O. Gutiérrez. "Efficient ray-tracing techniques for three-dimensional analyses of propagation in mobile communications: Application to picocell ands microcell scenarios." *IEEE Antennas and Propagation Magazine*, 40(2):15–28, April 1998.

[36] C. Tornevik, J. E. Berg, and F. Lotse. "900 MHz propagation measurements and path loss models for different indoor environments." *Proceedings of. IEEE VTC93*, 1993.

[37] S. Y. Seidel and T. S. Rappaport. "914 MHz path loss prediction models for indoor wireless communications in multifloored buildings." *IEEE Transactions on Antennas and Propagation*, 40(2):207–217, February 1992.

[38] J. B. Andersen, T. S. Rappaport, and S. Yoshida. "Propagation measurements and models for wireless communications channels." *IEEE Communications Magazine*, 33(1):42–49, January 1995.

[39] F. Saez de Adana, O. Gutiérrez, I. González, J. Pérez, and M. F. Cátedra. "Propagation model based on ray tracing for the design of personal communication systems in indoor environments." *IEEE Transactions on Vehicular Technology*, 49(6):2105–2112, November 2000.

[40] M. F. Cátedra and J. Pérez Arriaga. *Cell Planning for Wireless Communications*. Artech House, 1999.

[41] O. Gutiérrez, I. González, F. Saez de Adana, and M. F. Cátedra. "Application of GTD/UTD method to the analysis of the SPORT antenna on board the ISS." *IEEE Antennas and Propagation Society International Symposium 2000*, pp. 538–541, 2000.

[42] L. Lozano, M. J. Algar, I. González, and M. F. Cátedra. "FASANT: A versatile tool to analyze antennas and propagation in complex environments." *3rd European Conference on Antennas and Propagation 2009*, pp. 2088–2092, 2009.

[43] C. A. Balanis. *Advanced Engineering Electromagnetics*. John Wiley and Sons, 1989.

[44] D. A. McNamara, C. W. I. Pistorius, and J. A. G. Malherbe. *Introduction to the Uniform Geometrical Theory of Diffraction*. Artech House, 1990.

[45] R. G. Kouyoumjian. "Asymptotic high-frequency methods." *Proceedings of the IEEE*, 53(8):864–876, August 1965.

[46] G. A. Deschamps. "Ray techniques in electromagnetics." *Proceedings of the IEEE.*, 6(9):1022–1035, September 1972.

[47] W. H. Press, B. P. Flannery, S. A. Teukolsky, and W. T. Vetterling. *Numerical Recipes*. Cambridge University Press, 1987.

[48] P. H. Pathak. "An asymptotic analysis of the scattering of plane waves by a smooth convex cylinder." *Radio Science*, 14:419–435, May-June 1979.

[49] L. P. Eisenhart. *A Treatise on the Differential Geometry of Curves and Surfaces*. Dover Publications, Inc., 1960.

Chapter 4

Physical Optics and Physical Theory of Diffraction (PO/PTD)

4.1 INTRODUCTION TO PO/PTD

Physical optics (PO) is an intermediate method, between geometrical optics (GO) and full-wave electromagnetics [1, 2]. The PO method involves integrating the currents obtained from GO using the exact integral representation of the scattered fields. This approach assumes that the current on a surface is only dependent on the incident field and the geometrical parameters of each point on the surface [3, 4]. Therefore, this current is the same that would be obtained on an infinite plane tangent to the real surface at the point under analysis. The current is considered null in the shadowed areas of the surface, and the scattered field is obtained from the exact integral representation of that field as a function of the induced currents. These approximations make PO more exact than GO and also eliminate the problem of caustics. On the other hand, its computational requirements are higher. Another important feature of PO is that the obtained predictions are degraded when the direction of observation is far from the specular direction. The reason is that PO does not account for the contributions of the edges. An improvement in the formulation, called the physical theory of diffraction (PTD) [5], has been developed to avoid this problem. The PTD adds a nonuniform current term on the edges, introducing an additional contribution describing diffraction.

Historically, PO has also been known as the Kirchhoff approximation, the tangent plane approximation, or Huygen's principle [3], due to the approximations used to obtain the PO currents. However, both the Kirchhoff approximation and Huygen's principle are based on a scalar representation, as they were stated before

Maxwell's equations demonstrated the vector character of the fields. Therefore, these terms can only be applied to cases in which the scalar formulation is in agreement with the vector representation.

One of the problems in which the PO has been widely used is the RCS computation of electrically large targets. The necessity of using an asymptotic technique at certain frequencies makes the PO one of the best options available and much more suitable than GO/UTD because of the plane wave electromagnetic source used in the RCS computation. In this case, the problem of caustics appears very often in some types of targets. More specifically, if a plane wave illuminates a planar surface or a ruled surface (a surface which is straight in one of the parametric coordinates), GO/UTD is not able to predict the scattered field because a caustic arises. For this reason, most of the commercial software for computing RCS using asymptotic techniques is based on the PO approach.

Some of these software packages deserve mention: GRECO [6], which uses PO, PTD, the equivalent current method (ECM) [7], and the impedance boundary condition (IBC) [8]; CADDSCAT [9], which uses a PO-based integration technique consisting of analytical integration using bicubic patches over parametric surfaces and allows the computation of the RCS of targets covered by radar absorbing materials (RAM); XPATCH [10], which was initially based on the shooting and bouncing rays (SBR) technique [11] and PO, and more recently was improved to analyze realistic targets; FESS [12], which uses the method known as PTD-TC, where TC represents the transition currents obtained from the solutions to canonical problems; and POGCROS [13], which stems from the previously existing programs RANURS [14] and FASCRO [15] and is based on computing the PO integral using Gordon's method [16] or the stationary phase method (SPM) [17, 18], depending on whether the surface is planar or curved. POCGROS has grown extensively in recent years to include the computation of the RCS of targets covered by RAM as well as the contribution of multiple interactions to the scattered field. Another important feature of this code is the combination of PO with ray-tracing acceleration techniques to improve the computational efficiency of the RCS computation.

Most of these programs represent the targets by means of NURBS surfaces, to maintain compatibility with most of the CAGD tools available in the market and commonly used in the aeronautic, automobile, and ship industries. For this reason, a large number of relevant works have appeared in recent years, which have improved the PO formulation in order to apply it to bodies modeled by NURBS surfaces. Some contributions should be mentioned, such as the algorithm known as fast physical optics (FPO) [19, 20], in which the target is divided into subdomains and the scattered field is computed in each subdomain. After that, a phase correction is performed to obtain the scattered field of the entire target, or a parallelized Ludwig computation of the PO integral [21] is employed to reduce the computational time required to analyze a real target through PO.

Although the RCS computation has been the primary application of PO in electromagnetic analysis, this technique has also been used to solve other electromagnetic problems, such as the radiation problem. Some references should be mentioned, in which the radiation pattern of paraboloid reflector antennas [22] (a kind of configuration in which GO/UTD often encounters the problem of caustics, pyramidal horns [23], or impressed wire antennas are calculated [24]. Shifflet [25] applied PO to predict the effect of the radomes of antennas on their radiation patterns. However, the use of PO in the analysis of on-board antennas (which is also another radiation problem) is less common. As mentioned in Chapter 3, GO/UTD is preferred for this application because in most cases the results are accurate enough and caustics do not appear.

However, PO combined with the SPM can produce good predictions of the radiation patterns of on-board antennas over bodies modeled by NURBS surfaces, as shown in [26]. These results can be as precise as GO/UTD results, as proved in [27]. Therefore, the technique presented in this chapter can be used for this type of application in addition to the RCS computation. Therefore, this chapter will cover the PO formulation and will include the computation of the PO integral for both the RCS computation and the radiation problem.

4.2 PHYSICAL OPTICS FOUNDATIONS

The Stratton-Chu integral equations are a good starting point to deriving the PO expressions [28]. Those equations are obtained directly from Maxwell's equations by applying Green's theorem in its vector form [28, 29]. The Stratton-Chu equations give expressions for the scattered fields $\vec{E}_S(\vec{r})$, $\vec{H}_S(\vec{r})$ from a body described by a surface S at any point in free space with position vector $\vec{r}$, as a function of the tangential and normal components of the fields at the surface of the body:

$$\begin{aligned}\vec{E}_S(\vec{r}) &= \frac{1}{4\pi}\oint_S \left\{-j\omega\mu\left[\hat{n}\times\vec{H}(\vec{r}')\right]G(\vec{r},\vec{r}') + \left[\hat{n}\times\vec{E}(\vec{r}')\right]\times\nabla' G(\vec{r},\vec{r}')\right.\\ &+ \left.\left[\hat{n}\cdot\vec{E}(\vec{r}')\right]\times\nabla' G(\vec{r},\vec{r}')\right\}ds' \end{aligned} \tag{4.1}$$

$$
\begin{aligned}
\vec{H}_S(\vec{r}) &= \frac{1}{4\pi}\oint_S \Big\{ j\omega\epsilon \left[\hat{n}\times\vec{E}(\vec{r}')\right] G(\vec{r},\vec{r}') + \left[\hat{n}\times\vec{H}(\vec{r}')\right]\times\nabla' G(\vec{r},\vec{r}') \\
&+ \left[\hat{n}\cdot\vec{H}(\vec{r}')\right]\times\nabla' G(\vec{r},\vec{r}') \Big\}\, ds'
\end{aligned} \tag{4.2}
$$

where $\vec{r}'$ represents the position vector at a point on the surface, $\hat{n}$ is the unit normal vector at that point (see Figure 4.1), μ and ϵ represent the magnetic permeability and the electric permittivity of the vacuum, respectively, and ω is the angular frequency. G and $\nabla' G$ are Green's function and its gradient with respect to the point $\vec{r}'$. The Green function is defined as

$$
G(\vec{r},\vec{r}') = \frac{e^{-jk|\vec{r}-\vec{r}'|}}{|\vec{r}-\vec{r}'|} \tag{4.3}
$$

where k is the wavenumber.

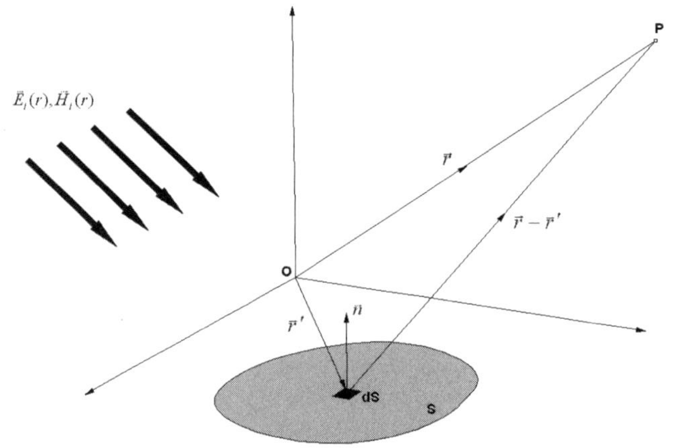

Figure 4.1 Scattering problem.

$\vec{E}(\vec{r}')$ and $\vec{H}(\vec{r}')$ are the total electric and magnetic fields on the body's surface, respectively; that is to say, they represent the sum of the incident and scattered fields:

$$
\vec{E}(\vec{r}') = \vec{E}_i(\vec{r}') + \vec{E}_s(\vec{r}') \tag{4.4}
$$

$$
\vec{H}(\vec{r}') = \vec{H}_i(\vec{r}') + \vec{H}_s(\vec{r}') \tag{4.5}
$$

The unknowns $\vec{E}(\vec{r}')$ and $\vec{H}(\vec{r}')$ appear on both sides of (4.1) and (4.2). Therefore, except in some particular cases, these two integral equations can only be solved with approximation techniques.

PO is one of these approximation methods. It is based on three assumptions that allow the Stratton-Chu integral equations to be simplified, reducing them to two integrals defined over the illuminated surface of the body [3, 4, 30, 31]. These assumptions are:

1. The field is null on the shadow areas of the surface.
2. The far-field approximation.
3. The tangent plane approximation.

The first assumption deals with the difficulty of obtaining the fields in the non-illuminated areas of the surface by assuming that they are null. This simplification is valid when the electrical size of the bodies is very large in comparison to the wavelength of the incident signal. As a result, the integrals are only defined over the illuminated parts of the body (see Figure 4.2).

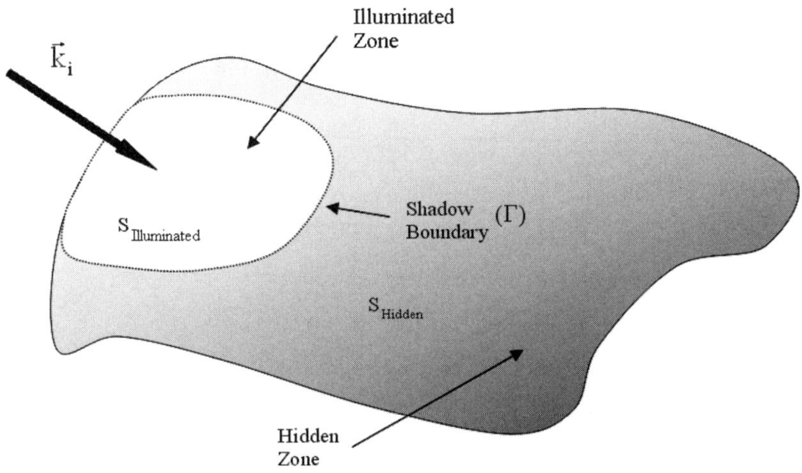

Figure 4.2 Illuminated surface and shadow boundary.

As a consequence of this assumption, the fields on the surface will be discontinuous at the curves separating the illuminated and shadow regions. This result violates one of the requirements for applying Green's theorem, and the resulting fields do not satisfy Maxwell's equations. These two problems can be solved by including a term in the expressions (4.1) and (4.2) to account for the discontinuity [3, 28, 29]. This term is a line integral along the shadow boundary. With this term, the fields can be written as:

$$\begin{aligned}\vec{E}_S(\vec{r}) &= \frac{1}{4\pi}\int_{S_i}\Big\{-j\omega\mu\left[\hat{n}\times\vec{H}(\vec{r}')\right]G(\vec{r},\vec{r}')+\left[\hat{n}\times\vec{E}(\vec{r}')\right]\times\nabla' G(\vec{r},\vec{r}')\\ &+ \left[\hat{n}\cdot\vec{E}(\vec{r}')\right]\times\nabla' G(\vec{r},\vec{r}')\Big\}\,ds' - \frac{1}{4\pi j\omega\epsilon}\oint_{\Gamma}(\nabla' G)\vec{H}(\vec{r}')d\vec{l} \qquad (4.6)\end{aligned}$$

$$\begin{aligned}\vec{H}_S(\vec{r}) &= \frac{1}{4\pi}\int_{S_i}\Big\{j\omega\epsilon\left[\hat{n}\times\vec{E}(\vec{r}')\right]G(\vec{r},\vec{r}')+\left[\hat{n}\times\vec{H}(\vec{r}')\right]\times\nabla' G(\vec{r},\vec{r}')\\ &+ \left[\hat{n}\cdot\vec{H}(\vec{r}')\right]\times\nabla' G(\vec{r},\vec{r}')\Big\}\,ds' - \frac{1}{4\pi j\omega\mu}\oint_{\Gamma}(\nabla' G)\vec{E}(\vec{r}')d\vec{l} \qquad (4.7)\end{aligned}$$

where Γ represents the curve defining the boundary between the illuminated and shadowed regions of the surface.

The second simplification consists of assuming that the observation point is in the far field with respect to the body (i.e., that the following condition is met):

$$|\vec{r}-\vec{r}'| > \frac{2D^2}{\lambda} \qquad (4.8)$$

where D is the maximum dimension of the body projected onto the plane tangent to the direction of incidence and λ is the wavelength. In this case, the amplitude term in the Green's function can be approximated by

$$|\vec{r}-\vec{r}'| = |\vec{r}| = r \qquad (4.9)$$

and the phase term by

$$|\vec{r}-\vec{r}'| = |\vec{r}| - \hat{r}\cdot\vec{r}' \qquad (4.10)$$

With this assumption, the Green's function and its gradient are

$$G(\vec{r},\vec{r}') = \frac{e^{-jkr}}{4\pi r}e^{-jk\hat{k}_s\cdot\vec{r}'} \qquad (4.11)$$

$$\nabla' G(\vec{r},\vec{r}') = \frac{e^{-jkr}}{4\pi r} jk\hat{k}_s e^{-jk\hat{k}_s \cdot \vec{r}'} \tag{4.12}$$

where $\hat{k}_s$ is the direction of scattering or observation.

The condition (4.8) can be considerably relaxed when the surface of the object is curved. In that case, D can be considered the maximal dimension of the first Fresnel zone over the surface of the body [3].

Introducing the expressions of the Green's function and its gradient into (4.6) and (4.7), the line integrals are transformed into surface integrals [3], giving:

$$\begin{aligned} \vec{E}_s(\vec{r}) &= \frac{j\omega\mu}{4\pi}\frac{e^{-jkr}}{4\pi r}\int_{S_i}\left\{-\left[\hat{n}\times\vec{H}(\vec{r}')\right]+\sqrt{\frac{\epsilon}{\mu}}\left[\hat{n}\times\vec{E}(\vec{r}')\right]\times\hat{k}_s\right. \\ &+ \left.\left(\hat{k}_s\cdot\left[\hat{n}\cdot\vec{H}(\vec{r}')\right]\right)\hat{k}_s\right\}e^{-j\vec{k}_s\cdot\vec{r}'}ds' \end{aligned} \tag{4.13}$$

$$\begin{aligned} \vec{H}_s(\vec{r}) &= \frac{j\omega\mu}{4\pi}\frac{e^{-jkr}}{4\pi r}\int_{S_i}\left\{\left[\hat{n}\times\vec{E}(\vec{r}')\right]-\sqrt{\frac{\epsilon}{\mu}}\left[\hat{n}\times\vec{H}(\vec{r}')\right]\times\hat{k}_s\right. \\ &+ \left.\left(\hat{k}_s\cdot\left[\hat{n}\cdot\vec{E}(\vec{r}')\right]\right)\hat{k}_s\right\}e^{-j\vec{k}_s\cdot\vec{r}'}ds' \end{aligned} \tag{4.14}$$

where $\vec{k}_s = k\hat{k}_s$.

These expressions satisfy Maxwell's equations and account for the possibility of edges on the body.

The third simplification is the assumption that the field over the surface of the body is a function of the incident field and the geometry of the surface at each point. In general, this is true when the principal radii of curvature at the points of the surface are larger than the wavelength. In this case, the surface in the vicinity of each point can be approximated by the plane tangent to the surface at that point. Therefore, the reflected field can be obtained from the incident wave, as the reflection of a plane wave from an infinite plane. This assumption allows the PO expressions to be described as a simple integral defined over the illuminated surface of the body because the fields on the surface satisfy:

$$\hat{n}\times\vec{E}(\vec{r}') = 0 \tag{4.15}$$

$$\hat{n}\times\vec{H}(\vec{r}') = 2\hat{n}\times\vec{H}_i(\vec{r}') = \vec{J}_S(\vec{r}') \tag{4.16}$$

where $\vec{J}_S(\vec{r}')$ is the current density on the surface.

Applying these boundary conditions to (4.13) and (4.14), the following expressions are obtained:

$$\vec{E}_s(\vec{r}) = \frac{j\omega\mu}{4\pi}\frac{e^{-jkr}}{4\pi r}\int_{S_i}\left\{-\left[\hat{n}\times\vec{H}(\vec{r}')\right]+\left(\hat{k}_s\cdot\left[\hat{n}\cdot\vec{H}(\vec{r}')\right]\right)\hat{k}_s\right\}e^{-j\vec{k}_s\cdot\vec{r}'}ds' \tag{4.17}$$

$$\vec{H}_s(\vec{r}) = \frac{jk}{2\pi}\frac{e^{-jkr}}{4\pi r}\int_{S_i}\left\{-\left[\hat{n}\times\vec{H}(\vec{r}')\right]\times\hat{k}_s+\right\}e^{-j\vec{k}_s\cdot\vec{r}'}ds' \tag{4.18}$$

Both expressions are valid for obtaining the scattered field in the far zone, but only the first one will be analyzed, to obtain an expression for the electric field.

Introducing the current density into (4.17), the electric field is given by

$$\vec{E}_s(\vec{r}) = \frac{-j\omega\mu}{4\pi}\frac{e^{-jkr}}{4\pi r}\int_{S_i}\left\{\vec{J}_S(\vec{r}')-\hat{k}_s\cdot\left(\hat{k}_s\cdot\vec{J}_s\right)\right\}e^{-j\vec{k}_s\cdot\vec{r}'}ds' \tag{4.19}$$

With the following property of vectors:

$$\vec{A}\times\vec{B}\times\vec{C} = \left(\vec{A}\cdot\vec{C}\right)\vec{B}-\left(\vec{A}\cdot\vec{B}\right)\vec{C} \tag{4.20}$$

the previous expression can be written as

$$\vec{E}_s(\vec{r}) = \frac{-j\omega\mu}{4\pi}\frac{e^{-jkr}}{4\pi r}\int_{S_i}\left\{\left(\hat{k}_s\times\vec{J}_s\right)\times\hat{k}_s\right\}e^{-j\vec{k}_s\cdot\vec{r}'}ds' \tag{4.21}$$

The cases of an incident plane wave and an incident spherical wave will both be considered. The first one corresponds to scattering problems, in which the objective is the computation of the RCS, and the second one corresponds to radiation problems, for the analysis of on-board antennas.

If the incident wave is planar, the incident field has the following form:

$$\vec{E}_i = \vec{E}_0 e^{-jk\hat{k}_i\cdot\vec{r}'} = \vec{E}_0 e^{-j\vec{k}_i\cdot\vec{r}'} \tag{4.22}$$

where $\vec{k}_i$ is the direction of incidence.

Therefore, the current density on the surface can be written as:

$$\vec{J}_s(\vec{r}') = \frac{2}{\eta}\left(\vec{E}_i\times\hat{k}_i\right)\times\hat{n} = \frac{2}{\eta}e^{-j\vec{k}_i\cdot\vec{r}'}\left(\vec{E}_0\times\hat{k}_i\right)\times\hat{n} \tag{4.23}$$

Replacing the current density in (4.21), the scattered field is

$$\vec{E}_s(\vec{r}) = \frac{-j}{\lambda}\frac{e^{-jkr}}{4\pi r}\int_{S_i}\left\{\left[\hat{k}_s \times \left(\left(\vec{E}_0 \times \hat{k}_i\right) \times \hat{n}\right)\right] \times \hat{k}_s\right\} e^{-j\vec{k}_s\cdot\vec{r}} ds' \quad (4.24)$$

In the RCS computation, the analysis is usually performed under conditions of monostatic radar, in which the incident plane wave arrives from the same direction as the scattered field. In that case, $\hat{k}_i = -\hat{k}_s$, considerably simplifying (4.24)

$$\vec{E}_s(\vec{r}) = \frac{-j}{\lambda}\frac{e^{-jkr}}{4\pi r}\vec{E}_0\int_{S_i}\left\{\left(\hat{k}_s \cdot \hat{n}\right) e^{2j\vec{k}_s\cdot\vec{r}'}\right\} ds' = \frac{-j}{\lambda}\frac{e^{-jkr}}{4\pi r}\vec{E}_0 I_{RCS} \quad (4.25)$$

where I_{RCS} is the PO integral for the RCS computation.

In this case, the polarization is conserved; that is to say, $\vec{E}_s$ has the same direction as $\vec{E}_0$.

If the incident wave is spherical, the current density must be written as a function of the incident magnetic field, in the following way:

$$\vec{J}_s(\vec{r}') = 2\hat{n} \times \vec{H}_0(\vec{r}')e^{-jk|\vec{k}_i|} = 2\hat{n} \times \vec{H}_0(\vec{r}')e^{-jkd} \quad (4.26)$$

where $\vec{H}_0(\vec{r}')$ is the amplitude of the magnetic field at the point $\vec{r}'$ and d is the distance from the source to that point.

Therefore, the scattered field for this case is written as:

$$\begin{aligned}\vec{E}_s(\vec{r}) &= \frac{-j\omega\mu}{2\pi}\frac{e^{-jkr}}{4\pi r}\int_{S_i}\left\{\hat{k}_s \times \left[\hat{n} \times \vec{H}_0(\vec{r}')\right] \times \hat{k}_s\right\} e^{-jk\left(\hat{k}_s\cdot\vec{r}'-d\right)} ds' \\ &= \frac{-j\omega\mu}{2\pi}\frac{e^{-jkr}}{4\pi r} I_{RP} \quad (4.27)\end{aligned}$$

where I_{RP} is the PO integral for radiation problems.

4.3 METHODS TO CALCULATE THE PO INTEGRAL FOR COMPLEX BODIES

4.3.1 Numerical Integration

There are several techniques for the numerical computation of integrals: the trapeze method, Simpson's method, Romney's method, Gaussian methods, and so on [32].

All of them are based on approximating the integrand function by an easily integrable function within the integration domain. The approximated function is obtained from a set of discrete values of the original function (samples) in the integration domain. The number of samples considered in the approximated function must be enough to accurately represent the real function, so that the value of the integral of the approximated function will be close to the real value of the integral.

The application of numerical techniques to computing the PO integral can present some problems when k is large compared to the surface to integrate (i.e., the wavelength of the incident signal λ is small in comparison with the size of the body). In this situation, the phase of the integrand varies rapidly along the surface, and therefore the number of samples needed to obtain the approximated function is very large. The minimum number of samples to obtain accurate results is considered to be 10 times the size of the surface in units of the wavelength.

Therefore, numerical integration can be applied to electrically small surfaces. Because PO is an asymptotic technique, valid in the case of electrically large objects, the evaluation of the PO integral using numerical techniques is not the optimal way to proceed, although numerical integration is sometimes used to validate other integration techniques.

4.3.2 Ludwig Method

The Ludwig method is used for far-field computations when double integrals are involved [33]. The main advantage of this approach is that it substantially reduces the number of points required to numerically evaluate the integral. To apply this method, the PO integral is written over a parametric surface:

$$I_{RCS,RP} = \int_0^1 \int_0^1 g(u,v) e^{jkf(u,v)} du dv \tag{4.28}$$

where u and v are the parametric coordinates of the surface.

The functions $g(u,v)$ and $f(u,v)$ are, respectively:

$$g(u,v) = \hat{k}_s \cdot \hat{n} \tag{4.29}$$

$$f(u,v) = 2\hat{k}_s \cdot \vec{r}' \tag{4.30}$$

for the RCS computation and

$$g(u,v) = \left\{ \hat{k}_s \times \left[\hat{n} \times \vec{H}_0(\vec{r}') \right] \times \hat{k}_s \right\} \tag{4.31}$$

$$f(u,v) = \hat{k}_s \cdot \vec{r}' - d \tag{4.32}$$

for the radiation problem.

Consider an incremental area of S called ΔS_{mn} in which the dimensions are on the order of a wavelength and thus the phase value cannot vary more than 2π, meaning that the integrand will vary slowly over ΔS_{mn}. However, the 2π phase variation could cause the integrand to behave like a full cycle of a sinusoid, which still needs a large number of points to approximate the entire integrand and numerically evaluate the integral. However, the amplitude and phase functions can be approximated individually. In this case, a simple linear form will be sufficient over ΔS_{mn}. The expressions for the amplitude and phase functions can be written as:

$$g(u,v) \approx a_{mn} + b_{mn}(u - u_m) + c_{mn}(u - u_n) \tag{4.33}$$

$$f(u,v) \approx \alpha_{mn} + \beta_{mn}(u - u_m) + \gamma_{mn}(u - u_n) \tag{4.34}$$

The parameters a_{mn}, b_{mn}, c_{mn}, α_{mn}, β_{mn}, and γ_{mn} can be obtained by minimizing the mean-squared error [32].

The integration over ΔS_{mn} can be performed analytically from expressions (4.33) and (4.34), and its contribution to the integral is:

$$\begin{aligned} \Delta I_{mn} &= e^{jk\alpha_{mn}} \Bigg\{ a_{mn} \left[\frac{e^{jk\beta_{mn}\Delta\theta_m} - 1}{jk\beta_{mn}} \right] \left[\frac{e^{jk\gamma_{mn}\Delta\phi_n} - 1}{jk\gamma_{mn}} \right] \\ &+ b_{mn} \left[\frac{\Delta\theta_m}{jk\beta_{mn}} e^{jk\beta_{mn}\Delta\theta_m} - \frac{e^{jk\beta_{mn}\Delta\theta_m} - 1}{(jk\beta_{mn})^2} \right] \left[\frac{e^{jk\gamma_{mn}\Delta\phi_n} - 1}{jk\gamma_{mn}} \right] \\ &+ c_{mn} \left[\frac{e^{jk\beta_{mn}\Delta\theta_m} - 1}{jk\beta_{mn}} \right] \left[\frac{\Delta\phi_n}{jk\gamma_{mn}} e^{jk\gamma_{mn}\Delta\phi_n} - \frac{e^{jk\gamma_{mn}\Delta\phi_n} - 1}{(jk\gamma_{mn})^2} \right] \Bigg\} \end{aligned} \tag{4.35}$$

The PO integral is then obtained by summing these contributions over m and n.

4.3.3 Method of Plane Facets

This method approximates the object by a set of plane facets. It was mentioned in Chapter 2 that a complex body can be represented by a faceted version of the real object. This method has been widely used [34, 35] to calculate the RCS because it allows the PO integral to be evaluated analytically using Gordon's method [16].

The PO integral for the RCS computation on a plane facet is simplified because the normal vector $\hat{n}$ is constant over the entire surface. Therefore:

$$I_{RCS} = \hat{k}_s \cdot \hat{n} \int_S e^{2j\vec{k}_s \cdot \vec{r}'} ds' \tag{4.36}$$

Considering a Cartesian axis system in which the Z axis is parallel to the vector normal to the surface (see Figure 4.3), the previous expression can be rewritten as:

$$I_{RCS} = \hat{k}_s \cdot \hat{z} \int_S e^{2jk(k_x x' + k_y y')} ds' \tag{4.37}$$

where

$$\hat{k}_s = k_x \hat{x} + k_y \hat{y} + k_z \hat{z} \tag{4.38}$$

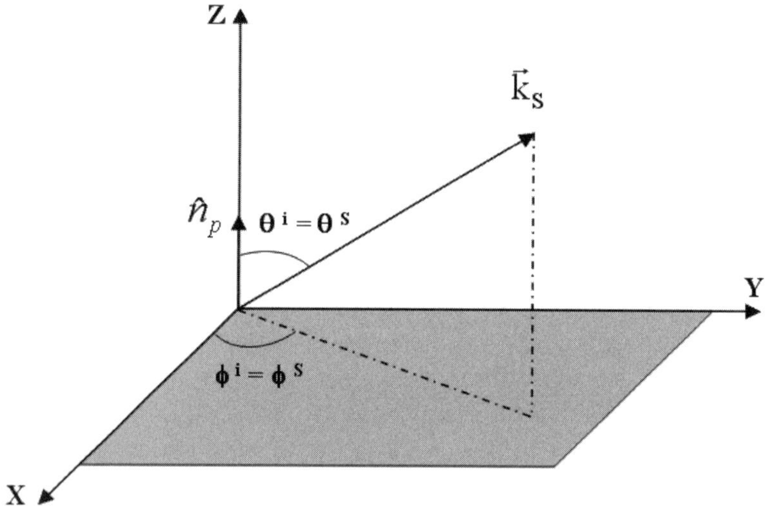

Figure 4.3 System of reference for the method of plane facets.

Using Gordon's method, this double integral can be expressed as a one-dimensional integral along the contour of the surface:

$$I_{RCS} = \begin{cases} \hat{k}_s \cdot \hat{z} \oint_{\gamma} \dfrac{-1}{2jkk_x} e^{2jk(k_x x' + k_y y')} \hat{y} d\vec{l} & ;\text{if } k_x \neq 0 \\ \hat{k}_s \cdot \hat{z} \oint_{\gamma} \dfrac{-1}{2jkk_y} e^{2jk(k_x x' + k_y y')} \hat{x} d\vec{l} & ;\text{if } k_y \neq 0 \\ \hat{k}_s \cdot \hat{z} \cdot AREA & ;\text{if } k_x = k_y = 0 \end{cases} \tag{4.39}$$

where $AREA$ is the area of the planar surface on which the integral is calculated.

If the surface is polygonal, the previous expressions can be calculated analytically, giving:

$$I_{RCS} = \begin{cases} \hat{k}_s \cdot \hat{z} \displaystyle\sum_{n=1}^{N} \hat{y} \Delta\vec{a}_n \frac{1}{kk_x} SINC\left[\frac{k\hat{k}\Delta\vec{a}_n}{\pi}\right] e^{\frac{j}{2}k\hat{k}\left(\vec{a}_{n+1} + \vec{a}_n\right)} & ;\text{if } k_x \neq 0 \\ \hat{k}_s \cdot \hat{z} \displaystyle\sum_{n=1}^{N} \hat{x} \Delta\vec{a}_n \frac{1}{kk_y} SINC\left[\frac{k\hat{k}\Delta\vec{a}_n}{\pi}\right] e^{\frac{j}{2}k\hat{k}\left(\vec{a}_{n+1} + \vec{a}_n\right)} & ;\text{if } k_y \neq 0 \\ \hat{k}_s \cdot \hat{z} \cdot AREA & ;\text{if } k_x = k_y = 0 \end{cases} \tag{4.40}$$

where N is the number of vertices of the polygon, $\vec{a}_n$ is the position vector of the Nth vertex and $\Delta\vec{a}_n = \vec{a}_{n+1} + \vec{a}_n$.

The area of the polygon and the unit normal vector can be calculated from the vertices of the polygon with the following expression:

$$2AREA \cdot \hat{n} = \vec{a}_1 \times \vec{a}_2 + \vec{a}_2 \times \vec{a}_3 + ... + \vec{a}_{N-1} \times \vec{a}_N + \vec{a}_N \times \vec{a}_1 \tag{4.41}$$

Modeling an object using plane facets introduces an inherent error when the integral is evaluated over curved surfaces. This error is due to the geometric approximation known as a faceting error. This error arises in the phase term of the integral because the points of the facet are slightly separated from the real surface of the body. The effect of this error is a small ripple in the field diagram around the correct values. This effect is larger when the frequency increases.

To reduce this error, experience has led to the conclusion that the real surface must be approximated by a faceted model with an error less than $\lambda/16$ [35]. Therefore, the necessary number of facets is very large in complex structures, and presents a number of associated disadvantages.

The faceted error can be considerably reduced by adding the contributions of the artificial edges of the model, formed by the connections between the plane facets and evaluated using PTD, to the PO contribution [34]. This procedure smooths the ripples and produces most accurate predictions. The PTD formulation will be described later in this chapter.

4.3.4 Method of Numerical Quadrature

The method of numerical quadrature has also been used in the calculation of the RCS of complex bodies [36]. The idea is to approximate the amplitude and phase terms of the PO integral by simple biquadratic functions to obtain analytically integrable expressions. In the RCS computation, the amplitude and phase terms of the integrand are approximated using parametric functions as follows:

$$\hat{k}_s \cdot \hat{n}(\vec{r}') \approx \sum_{i=1}^{3} \sum_{j=1}^{3} \phi_i(u)\phi_j(v) n_{ij} \tag{4.42}$$

$$2j\vec{k}_s \cdot \vec{r}' \approx 2jk \sum_{i=1}^{3} \sum_{j=1}^{3} \phi_i(u)\phi_j(v) p_{ij} \tag{4.43}$$

where

$$\phi_1(s) = 2s^2 - 3s + 1 \tag{4.44}$$

$$\phi_2(s) = -4s^2 + 4s \tag{4.45}$$

$$\phi_3(s) = 2s^2 - s \tag{4.46}$$

$$n_{ij} = \hat{k}_s \cdot \hat{n}_{ij} \tag{4.47}$$

$$p_{ij} = \hat{k}_s \cdot \vec{r}'_{ij} \tag{4.48}$$

where $\vec{r}'_{ij}$ are 9 sample points on the integration surface and $\hat{n}_{ij}$ are the normal vectors to the surface at the sample points (see Figure 4.4). With these approximations, the integral becomes:

$$I_{RCS} \approx \sum_{i=1}^{3} \sum_{j=1}^{3} n_{ij} \int_0^1 \int_0^1 \phi_i(u)\phi_j(v) e^{2jk \sum_{k=1}^{3} \sum_{l=1}^{3} \phi_k(u)\phi_l(v) p_{kl}} du dv \tag{4.49}$$

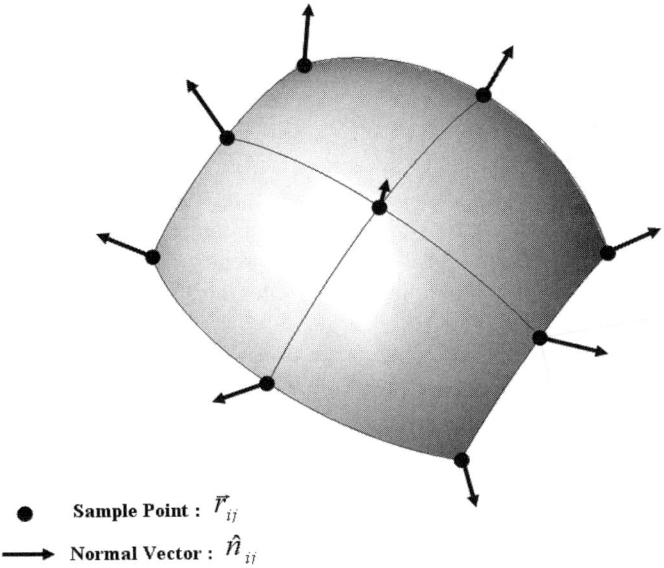

Figure 4.4 Sample points and vectors normal to the surface.

The previous expression cannot be analytically integrated; the phase term must be further approximated. In each of the integrals of (4.49), the amplitude term only has non negligible values in the vicinity of the sample points, due to the features of the functions $\phi_i(s)$. The phase term can be further approximated using this fact. For instance, in the first integral

$$I_{11} = \int_0^1 \int_0^1 \phi_1(u)\phi_1(v) e^{2jk \sum_{k=1}^{3} \sum_{l=1}^{3} \phi_k(u)\phi_l(v) p_{kl}} dudv \tag{4.50}$$

the amplitude term takes very small values at every point (u, v), except for points located in the vicinity of $(0, 0)$, which corresponds to the sample point $\vec{r}_{11}$. Therefore, the phase terms with degrees larger than 2 are neglected, and the phase of the integrand is approximated as follows:

$$2jk\sum_{k=1}^{3}\sum_{l=1}^{3}\phi_k(u)\phi_l(v)p_{kl} \approx 2jk\left(a_1u^2 + a_2v^2 + a_3uv + a_4u + a_5v + a_6\right) \tag{4.51}$$

One more approximation is necessary before this can be analytically integrated. This approximation consists of neglecting the cross term (i.e., setting $a_3 = 0$). In this way, the variables u and v are not coupled in either the amplitude or the phase terms, and the integral can be analytically evaluated by transforming the double integral into two uncoupled simple integrals:

$$\int \left(b_2x^2 + b_1x + b_0\right) e^{j\left(c_2x^2+c_1x+c_0\right)} = b_2I_2 + b_1I_1 + b_0I_0 \tag{4.52}$$

where

$$I_2 = \int x^2 e^{j\left(c_2x^2+c_1x+c_0\right)} \tag{4.53}$$

$$I_1 = \int x e^{j\left(c_2x^2+c_1x+c_0\right)} \tag{4.54}$$

$$I_0 = \int e^{j\left(c_2x^2+c_1x+c_0\right)} \tag{4.55}$$

These integrals can each be evaluated analytically. An analogous treatment can be performed on the rest of the integrals of (4.49).

As expected, this method produces good results, especially in the case of bicubical patches, although the results are still not exact for this type of surface.

This method can be considered a generalization of the method of plane facets because it shares several common properties:

1. It provides close expressions to the PO integral. However, these expressions are more complicated than in the case of the plane facets, and are inexact even for bicubical surfaces.

2. The geometry of the body is approximated when the integral is evaluated. This approximation is, usually, more approximate than in the method of plane facets.

3. It produces errors, due to the faceting inherent in the approximation. Therefore, this model is not valid for all frequencies. In general, the error is lower than

that of the method of plane facets, and thus the problems are not as important. However, in the plane facet method, the result can be improved by including the contribution of the PTD of the artificial edges, which is not possible in this method. In both methods, the results improve as the number of patches increases.

4.3.5 Method of the Graphical Processing of the Image

This method is also used for the computation of the RCS and is based on using the graphical hardware of a computer to obtain the geometrical information about the target necessary to compute the PO integral [6].

The idea is to obtain a 3-D image of the target representing only the illuminated part and eliminating the hidden areas, for a given direction of incidence (point of view). Moreover, it is possible to define the parameters of the source such that the basic colors RGB (red, green, blue) of each pixel coincide with the Cartesian coordinates of the unit vector, $\hat{n}$, normal to the surface of the body in the region represented by each pixel. With the position vector and the color components of each pixel, it is possible to evaluate the PO integral in an approximate way. In this method, the body is geometrically represented using NURBS surfaces.

The integral is evaluated in the following way: considering a Cartesian reference system in which the X and Y axes are contained in the plane of the screen and the Z axis is perpendicular to that plane, the PO integral for a given direction of incidence can be written as:

$$I_{RCS} = \int_S \cos\theta e^{2jkz} dS \tag{4.56}$$

where θ is the angle formed by the direction of incidence $\hat{k}$ and the normal vector of the surface at each point $\hat{n}$, and z is the distance between the point on the surface and the screen. S is the illuminated surface of the body, which coincides with the surface of the body represented on the screen. This is the projection of the real surface into the plane of the screen. Therefore, the differential element of surface area on the screen, dS', will be:

$$dS' = \cos\theta dS \tag{4.57}$$

where dS is the differential element of surface area on the object. Therefore, (4.56) can be written as

$$I_{RCS} = \int_A e^{2jkz} dS' \tag{4.58}$$

where A is the area of the screen. The available geometrical information is discretized using pixels, and the integral is therefore approximated as:

$$I_{RCS} = \sum_{pixels} e^{2jkz} \tag{4.59}$$

The previous expression is only valid if each pixel radiates as an infinitesimal aperture. In the regions of the body with incidence near grazing incidence ($\theta \rightarrow 90°$), an element of surface $\Delta S'$ can come from a very large element of surface ΔS. In that case, the assumption that each pixel radiates as an infinitesimal aperture does not hold. To solve the problem, it is assumed that each pixel is equivalent to a rectangular aperture with uniform illumination, and (4.59) can thus be written as:

$$I_{RCS} = \sum_{pixels} SINC\left(k\frac{l}{\cos\theta}\sin\theta\right) e^{2jkz} \tag{4.60}$$

where l is the size of the side of the aperture associated with each pixel projected on the screen.

4.3.6 Current Modes

The current modes approach states that the current in the smooth areas of electrically large bodies can be expressed accurately and efficiently in terms of complex exponential functions, also called modes [37]. These functions are characterized by amplitude and phase functions that vary slowly over large areas. Representing the PO current as current modes allows the efficient procedure presented in [38] to be used to evaluate the PO integral. The following procedure is the same whether the PO integral is to be evaluated for the RCS computation or the radiation problem.

The idea is to approximate the PO current $\vec{J}_S(\vec{r}')$ in the smooth parts of the body under analysis by a finite number of current modes, as follows:

$$\vec{J}_S(\vec{r}') = \sum_{m=1}^{N} \vec{J}_m(\vec{r}') e^{-j\varphi_m(\vec{r}')} \tag{4.61}$$

where $\vec{J}_m(\vec{r}')$ and $\varphi_m(\vec{r}')$ are the amplitude and phase functions of the Mth mode. The field at a given observation point due to this current mode is given by:

$$I = \int_S \vec{J}_m(\vec{r}')e^{-j\varphi_m(\vec{r}')}G(\vec{r},\vec{r}')dS \tag{4.62}$$

where G is the Green's function and S is the surface that supports the current mode. When the observation point is not very close to the mode (which happens in both problems analyzed in this chapter), the integral (4.61) over a NURBS surface can be written as:

$$I = \int_S \vec{S}(\vec{r}'(u,v))e^{-j\varphi_m(\vec{r}'(u,v))}dS \tag{4.63}$$

where the integral has been expressed in terms of an amplitude and a phase function, which include both the amplitude and phase corresponding to the current mode as well as the Green's function. Therefore, the integral in (4.63) can be considered a current mode because its amplitude and phase terms change slowly along the surface.

The phase term can be expressed using a Taylor expansion around the middle point (u_0, v_0) of the surface:

$$\Phi(u,v) = \Phi(u_0,v_0) + \omega_{u_0}(u-u_0) + \omega_{v_0}(v-v_0) + R(u,v) \tag{4.64}$$

where

$$\omega_{u_0} = \frac{\partial}{\partial u}\Phi(u,v)(u-u_0)\bigg|_{u_0,v_0} \tag{4.65}$$

$$\omega_{v_0} = \frac{\partial}{\partial v}\Phi(u,v)(u-u_0)\bigg|_{u_0,v_0} \tag{4.66}$$

and the residual function $R(u,v)$ includes the second- and higher-order terms of the Taylor series.

Considering the change of variables:

$$z = u - 0.5 \tag{4.67}$$

$$t = v - 0.5 \tag{4.68}$$

(4.63) can be rewritten as:

$$I = e^{-jA}\int_{-0.5}^{0.5}\int_{-0.5}^{0.5} \vec{S}'(z,t)e^{-j\omega_{u_0}z}e^{-j\omega_{v_0}t}dzdt \tag{4.69}$$

where

$$A = [\Phi(u_0, v_0) - \omega_{u_0} \cdot u_0 - \omega_{v_0} \cdot v_0 + 0.5\omega_{u_0} + 0.5\omega_{v_0}] \tag{4.70}$$

and

$$\vec{S}'(z,t) = \vec{S}(z + 0.5t + 0.5)e^{jRz+0.5t+0.5)} \tag{4.71}$$

The domain of the integral in (4.69) can be extended to $(-\infty, \infty)$ by including the corresponding pulse functions:

$$I = e^{-jA} \int_{-\infty}^{\infty} \int_{-\infty}^{\infty} \vec{S}'(z,t) e^{-j\omega_{u0} z} e^{-j\omega_{v0} t} \Pi(z)\Pi(t) dz dt \tag{4.72}$$

$$\Pi(x) = \begin{cases} 1 & \text{for } |x| \leq 0.5 \\ 0 & \text{elsewhere} \end{cases} \tag{4.73}$$

Expression (4.72) is the Fourier transform of the function $\vec{S}'(z,t)\Pi(z)\Pi(t)$ at the frequencies $(\omega_{u_0}\omega_{v_0})$ and can be accurately computed by expanding $\vec{S}'(z,t)$ in a polynomial form of degree $(M \times N)$ in the following way:

$$\vec{S}'(z,t) = \sum_{m=1}^{M} \sum_{n=1}^{N} \vec{b}_{mn} z^m t^n \tag{4.74}$$

Following classical procedures, the coefficients of the polynomial can be obtained by minimizing the squared mean error [32].

Using this property of the Fourier transform:

$$F\left[t^n x(t)\right] = j^n \frac{d^n}{d\omega^n} X(\omega) \tag{4.75}$$

the integral in (4.72) can be evaluated as:

$$I = e^{-jA} \sum_{m=1}^{M} \sum_{n=1}^{N} \vec{b}_{mn} j^m j^n \times \frac{d^n}{d\omega^n} \left[SINC\left(\frac{\omega u}{2\pi}\right) SINC\left(\frac{\omega v}{2\pi}\right) \right] \Big|_{w_{u0}, w_{v0}} \tag{4.76}$$

Alternatively, $\vec{S}'(z,t)$ can be approximated by a series of pulses

$$\vec{S}'(z,t) = \sum_{i=-N_u/2}^{N_u/2} \sum_{i=-N_v/2}^{N_v/2} \vec{S}'(z_i,t_j)\Pi\left(\frac{z-z_i}{N_u}\right) \times \Pi\left(\frac{t-t_j}{N_v}\right) \tag{4.77}$$

and the integral (4.69) would then be:

$$\begin{aligned} I &= e^{-jA}\left[\frac{1}{N_u}SINC\left(\frac{\omega_u}{2\pi N_u}\right)\frac{1}{N_v}SINC\left(\frac{\omega_v}{2\pi N_v}\right)\right] \\ &\times \sum_{i=-N_u/2}^{N_u/2} \sum_{i=-N_v/2}^{N_v/2} \vec{S}'(z_i,t_j)e^{-j\omega_{u0}z\omega_{v0}t} \end{aligned} \tag{4.78}$$

4.4 THE STATIONARY PHASE METHOD

The Stationary Phase Method (SPM) is a mathematical tool that allows for calculating integrals of the form:

$$I = \int_H g(\vec{x})e^{jkf(\vec{x}}dx_1dx_2...dx_n \qquad \vec{x} = (x_1, x_2, ..., x_n) \tag{4.79}$$

where H represents the integration domain, $\vec{x}$ is an N-dimensional variable defined on the integration domain, in this case, $\vec{x} = (u,v)$ is a two-dimensional variable formed by the two parametric coordinates, $g(\vec{x}) = g(u,v)$ and $f(\vec{x}) = f(u,v)$ are the amplitude and phase functions, respectively, which are analytical functions that vary slowly within H, and k is a large real number, which means that the integrand varies rapidly within the integration domain H.

The SPM can be applied to both the RCS computation and the radiation problem because the amplitude and phase functions are the same as in the Ludwig method, given in (4.29) and (4.30) for the RCS computation and in (4.31) and (4.32) for the radiation problem. The SPM for double integrals [17, 18] can be used to calculate the integral (4.79) on parametric surfaces. The value of the integral is approximated by three main contributions, due to three kinds of critical points:

1. Stationary phase points (or first order);
2. Boundary points (or second order);

3. Vertex points (or third order).

The contribution of each of these critical points can be expressed as an infinite sum of negative powers of k. Except in some particular cases, the main contributions from each type of point are:

1. Stationary phase points: $I_s \propto k^{-1} + O(k^{-2})$
2. Boundary points: $I_c \propto k^{-3/2} + O(k^{-5/2})$
3. Vertex points: $I_v \propto k^{-2} + O(k^{-3})$

Mathematically, stationary phase points satisfy:

$$f_u(u_s, v_s) = 0 \wedge f_v(u_s, v_s) = 0 \tag{4.80}$$

where $f_u(u, v)$ and $f_v(u, v)$ are the parametric derivatives of the phase function.

For the RCS computation, this condition is equivalent to:

$$\vec{k}_s \cdot \vec{r}_u'(u_s, v_s) = 0 \wedge \vec{k}_s \cdot \vec{r}_v'(u_s, v_s) = 0 \tag{4.81}$$

while for the radiation problem, this condition is:

$$\left(\vec{k}_s - \vec{k}_i\right) \cdot \vec{r}_u'(u_s, v_s) = 0 \wedge \left(\vec{k}_s - \vec{k}_i\right) \cdot \vec{r}_v'(u_s, v_s) = 0 \tag{4.82}$$

The stationary phase points are the specular points on the surface (see Figure 4.5).

Applying the SPM for double integrals, the contribution of these points is given by:

$$I_s = \frac{\pi g^s}{k} e^{jkf^s} \sqrt{\frac{1}{\left|f_{uu}^s f_{vv}^s - (f_{uv}^s)^2\right|}}\, e^{jk\frac{\pi}{4}\sigma(\delta+1)} \tag{4.83}$$

where $\sigma = sign(f_{vv})$ and $\delta = sign\left(f_{uu} f_{vv} - (f_{uv})^2\right)$.

If there is more than one stationary phase point on the surface, their contributions are added to the PO integral. If the specular points are very close to each other with respect to the wavelength, the expression (4.83) would not be rigorously valid because it assumes that the stationary phase points are isolated. However, these points are always isolated except in surfaces with abrupt changes in the normal vector, and PO is not applicable to these surfaces.

Other special cases where (4.83) is not applicable are:

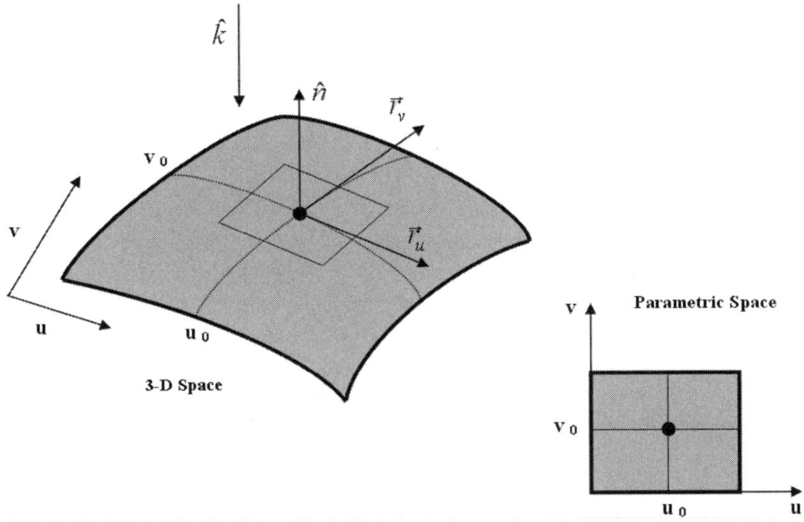

Figure 4.5 Stationary phase points.

- If the denominator of (4.83) is zero (degenerate points) (i.e., $f_{uu}f_{vv} = f_{uv}^2$). In this case, the contribution to the integral is proportional to $k^{-5/6}$ and is computed in the following way:

  (a) If $f_{uu} \neq 0$

$$I_s = g^s e^{jkf^s} \frac{2^{5/6}\sqrt{\pi}\sqrt{|f_{uu}^s|}\,\Gamma(\frac{1}{3})}{3^{1/6}|B|^{1/3}k^{5/6}} e^{jk\frac{\pi}{4}sign\,(f_{uu}^s)} \tag{4.84}$$

  where $B = (f_{uu}^s)^3 f_{vvv}^s - 3(f_{uu}^s)^2 f_{uv}^s f_{uvv}^s + 3f_{uu}^s (f_{uv}^s)^2 f_{uuv}^s - (f_{uv}^s)^3 f_{uuu}^s$.

  (b) If $f_{uu} = 0$

$$I_s = g^s e^{jkf^s} \frac{2^{5/6}\sqrt{\pi}\sqrt{|f_{vv}^s|}\,\Gamma(\frac{1}{3})}{3^{1/6}|B|^{1/3}k^{5/6}} e^{jk\frac{\pi}{4}sign\,(f_{vv}^s)} \tag{4.85}$$

  where $B = (f_{vv}^s)^3 f_{uuu}^s - 3(f_{vv}^s)^2 f_{uv}^s f_{uuv}^s + 3f_{vv}^s (f_{uv}^s)^2 f_{uvv}^s - (f_{uv}^s)^3 f_{vvv}^s$.

- If the stationary phase point coincides with one of the integration vertices; that is, $(u_s, v_s) = (0,0), (0,1), (1,0)$ or $(1,1)$. In this case, the contribution to the integral is computed in the following way:

(a) If it is a local minimum or maximum, $f_{uu}f_{vv} > f_{uv}^2$:

$$I_s = \frac{\pi g^s}{k} e^{jkf^s} \sqrt{\frac{1}{\left| f_{uu}^s f_{vv}^s - (f_{uv}^s)^2 \right|}} \tan^{-1}\left[\frac{\sqrt{f_{uu}^s f_{vv}^s - (f_{uv}^s)^2}}{f_{uv}^s} \right] \quad (4.86)$$

(b) If it is a saddle point, $f_{uu}f_{vv} < f_{uv}^2$:

$$I_s = \frac{\pi g^s}{2k} e^{jkf^s} \sqrt{\frac{1}{-f_{uu}^s f_{vv}^s}} \quad (4.87)$$

- If the stationary phase point is at a boundary, but not at a vertex. In this case, the value of I_s is half of (4.83).

  Mathematically, boundary points satisfy one of the following conditions:

$$f_u(u_c, 0) = 0,\, f_v(u_c, 1) = 0,\, f_v(0, v_c) = 0,\, f_v(1, v_c) = 0 \quad (4.88)$$

  For the RCS computation, these conditions are equivalent to:

$$\vec{k}_s \cdot \vec{r}'_u(u_c, 0) = 0, \vec{k}_s \cdot \vec{r}'_v(u_c, 1) = 0, \vec{k}_s \cdot \vec{r}'_v(0, v_c) = 0, \vec{k}_s \cdot \vec{r}'_v(1, v_c) = 0 \quad (4.89)$$

and for the radiation problem:

$$\begin{aligned} &\left(\vec{k}_s - \vec{k}_i\right) \cdot \vec{r}'_u(u_c, 0) = 0,\ \left(\vec{k}_s - \vec{k}_i\right) \cdot \vec{r}'_v(u_c, 1) = 0, \\ &\left(\vec{k}_s - \vec{k}_i\right) \cdot \vec{r}'_u(0, v_c) = 0,\ \left(\vec{k}_s - \vec{k}_i\right) \cdot \vec{r}'_u(1, v_c) = 0 \end{aligned} \quad (4.90)$$

As one of the coordinates is fixed to 0 or 1, these are points on the boundary of the parametric space. Then, the boundary points will be on the curves that define the boundary of the surface, and they are also points for which the wave vector and the tangent vector to the curve are perpendicular (see Figure 4.6), which correspond to the diffraction points.

The contributions of these points to the PO integral are given by:

$$I_c = j(-1)^{\alpha} \frac{g^c}{k} e^{jkf^c} \frac{1}{f_{\alpha}^c} \sqrt{\frac{2j\pi}{kf_{\beta\beta}^c}} \quad (4.91)$$

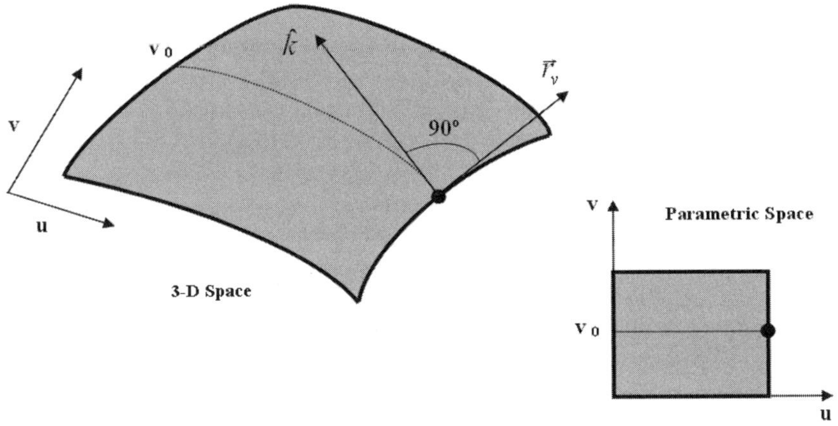

Figure 4.6 Boundary points.

where $\alpha \equiv u$ and $\beta \equiv v$ at the boundaries $u = 0$ and $u = 1$, and $\alpha \equiv v$ and $\beta \equiv u$ at the boundaries $v = 0$ and $v = 1$.

The previous expression is only valid if the boundary point is isolated from other boundary points and from stationary phase points. If the surface is smooth, the tangent vectors of the boundary curves will not change abruptly except at the vertices of the surfaces, and there will be no coupled boundary points. On the other hand, coupling between stationary phase points and boundary points is possible. The measurement of the coupling is given by the parameter:

$$\upsilon_\alpha = |f_\alpha^c| \sqrt{\frac{k}{2\,|f_{\alpha\alpha}^c|}} \tag{4.92}$$

evaluated at the boundary point. If $\upsilon_\alpha > 3$, the point is isolated and its contribution can be calculated with (4.91). Otherwise, the following expression must be used:

$$I_c = (-1)^\alpha \frac{2g^c}{k} e^{jkf^c} \sqrt{\frac{j\pi}{f_{\beta\beta}^c\,|f_{\alpha\alpha}^c|}}\, sign\left(f_\alpha^c\right) sF_s(\upsilon_\alpha) e^{-js\upsilon_\alpha^2} \tag{4.93}$$

where $sign\left(f_{\alpha\alpha}^c\right)$ and $F_s(x)$ is the Fresnel function of the argument x.

The previous expressions are uniform, in the sense that for large values of υ_α expression (4.93) tends to (4.91).

Vertex points are the four vertices of the integration domain, $(0,0)$, $(0,1)$, $(1,0)$, and $(1,1)$, which correspond to the four vertices of the surface (see Figure 4.7).

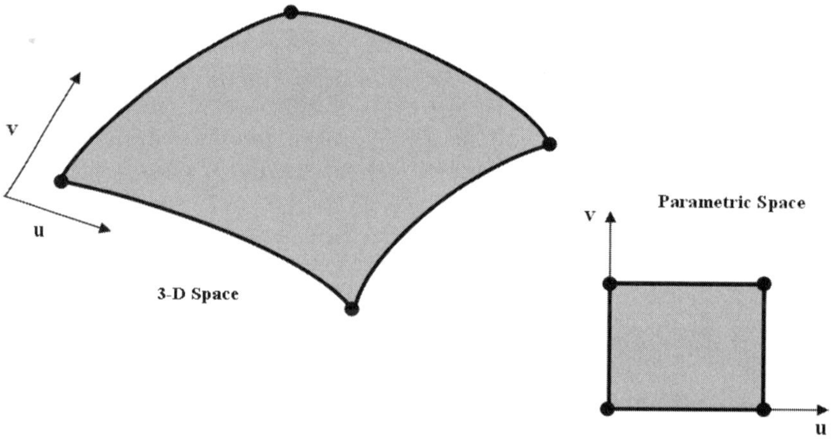

Figure 4.7 Vertex points.

If they are isolated from the stationary phase points and the boundary points, their contribution to the PO integral is given by

$$I_v = -g^v e^{jkf^v} \frac{(-1)^{u_v+v_v}}{k^2 f_u^v f_v^v} \tag{4.94}$$

If $\upsilon_u > 3$ or $\upsilon_v > 3$, the vertex point is not isolated and the previous expression must be replaced by:

$$I_v = (-1)^{u_v+v_v} \frac{2g^v}{k} e^{jkf^v} \sqrt{\frac{1}{f_{uu}^v f_{vv}^v}} sign\left(f_u^v\right) sign\left(f_v^v\right) T_u T_v \tag{4.95}$$

where

$$T_u = sign\left(f_{uu}^v\right) F_{sign(f_{uu}^v)}(\upsilon_u) e^{-j sign(f_{uu}^v) \upsilon_u^2} \tag{4.96}$$

and

$$T_v = sign\left(f_{vv}^v\right) F_{sign(f_{vv}^v)}(v_v) e^{-j sign(f_{vv}^v) v_v^2} \tag{4.97}$$

For larger values of v_u and v_v, the previous expression tends towards the contribution of an isolated point and, therefore, this is a uniform expression.

Obviously, to account for the contribution of any of these points (stationary phase, boundary, or vertex), it must be ensured that the point is illuminated by the source: an incident plane wave for the RCS computation or an antenna for the radiation problem.

The expressions shown above are only valid if $g(u, v)$ and $f(u, v)$ vary slowly in the domain of integration, but the integrand itself varies rapidly. These conditions are always satisfied in the radiation problem. However, for the RCS computation there are two situations in which this condition is not satisfied: on planar surfaces, the phase function will not vary rapidly enough for near-normal incidence. This is also the case if the surface is linear with respect to one of the parametric coordinates. For this reason, it is necessary to classify the patches before calculating the PO integral. The patches will be classified as:

1. Planar patches;
2. Singly curved patches that are linear with respect to one parametric coordinate;
3. Doubly curved patches.

The first two classes of patches cannot be directly treated using the SPM for double integrals. The solution to computing the PO integral on planar surfaces is very simple: apply Gordon's method, described in Section 4.3.3. For singly curved surfaces, it is necessary to perform a series expansion with respect to the linear coordinate, and then apply the SPM for the nonlinear coordinate. The procedure to calculate the PO integral for this kind of surface is explained next.

In the case of singly curved surfaces, $\vec{r}'(u, v)$ is linear in one of the parametric coordinates. This fact makes it possible to analytically calculate the integral for the linear variable and use the SPM for simple integrals [17, 18] for the nonlinear variable.

For instance, if the surface is straight in coordinate v, the phase term can be written as:

$$f(u, v) = 2\vec{k}_s \cdot \vec{r}_0'(u) + v 2\vec{k}_s \cdot \vec{r}_1'(u) = f_0(u) + v f_1(u) \tag{4.98}$$

and then the integral is

$$I = \int_{u=0}^{u=1} e^{jkf_0(u)} \int_{v=0}^{v=1} g(u,v)e^{vjkf_1(u)}dudv \tag{4.99}$$

The method to compute the integral I has two steps. The first is to integrate along the variable v, which can be done analytically. The function $I_v(u)$ is obtained, which depends only on u:

$$I_v(u) = e^{jkf_0(u)} \int_{v=0}^{v=1} g(u,v)e^{vjkf_1(u)}dv \tag{4.100}$$

The second step is integrate the previous function along the u coordinate:

$$I = \int_{u=0}^{u=1} I_v(u) \tag{4.101}$$

To obtain an expression for $I_v(u)$, the amplitude term of the integrand of expression (4.100), $g(u,v)$, is expanded in a Taylor's series about $v = 0.5$ (the mean point of the integration interval).

$$g(u,v) = g(u) + g^{(1)}(u)(v - v_0) + \frac{g^{(2)}(u)}{2!}(v - v_0)^2 + ... \tag{4.102}$$

where the functions $g^i(u)$ are the parametric derivatives of the function with respect to the v coordinate at the point $v = 0.5$:

$$g^{(i)}(u) = \frac{\partial^i g(u,v_0)}{\partial v^i} \tag{4.103}$$

If the function $g(u,v)$ varies smoothly, this function can be by the first terms of the Taylor series. Considering the first N terms of this series and grouping them by powers of v, (4.102) can be rewritten as:

$$g(u,v) = \sum_{n=0}^{N} \left(\sum_{k=n}^{N} \frac{g^k(u)(-v_0)^{k-n}}{(k-n)!n!} \right) v^n \tag{4.104}$$

The following expression is obtained by substituting expression (4.104) into (4.100):

$$I_v(u) \approx e^{jkf_0(u)} \sum_{n=0}^{N} \left(\sum_{k=n}^{N} \frac{g^k(u)(-v_0)^{k-n}}{(k-n)!n!} \right) A_n(u) \tag{4.105}$$

where

$$A_n(u) = \int_{v=0}^{v=1} v^n e^{vjkf_1(u)} dv \tag{4.106}$$

These integrals can be calculated analytically.

$$A_n(u) = \frac{n!}{[-jkf_1(u)]^{n+1}} - \sum_{i=0}^{n} \frac{n! e^{jkf_1(u)}}{(n-i)!\,[-jkf_1(u)]^{i+1}} \tag{4.107}$$

Therefore, a close expression for the function $I_v(u)$ has been obtained. Substituting (4.107) into (4.105), the following expression is obtained:

$$I_v(u) = I_{v1}(u) + I_{v2}(u) \tag{4.108}$$

where

$$I_{v1}(u) \approx e^{jkf_0(u)} \sum_{n=0}^{N} \frac{1}{[-jkf_1(u)]^{n+1}} \sum_{k=n}^{N} \frac{g^k(u)(-v_0)^{k-n}}{(k-n)!} \tag{4.109}$$

$$\begin{aligned} I_{v2}(u) \quad &\approx \quad e^{jk(f_0(u)+f_1(u))} \sum_{n=0}^{N} \left(\sum_{k=n}^{N} \frac{g^k(u)(-v_0)^{k-n}}{(k-n)!n!} \right) \\ &\cdot \left(\sum_{i=0}^{N} \frac{-1}{(k-n)!\,[-jkf_1(u)]^{i+1}} \right) \end{aligned} \tag{4.110}$$

The integral $I_v(u)$ can now be integrated along the u coordinate to evaluate I.

$$I = \int_{u=0}^{u=1} I_v(u) = \int_{u=0}^{u=1} I_{v1}(u) + \int_{u=0}^{u=1} I_{v2}(u) = I_1 + I_2 \tag{4.111}$$

Calling $G_1(u)$ and $G_2(u)$ the amplitude terms of the functions $I_{v1}(u)$ and $I_{v2}(u)$, respectively, the following is obtained:

$$I_1 = \int_{u=0}^{u=1} G_1(u) e^{jkf_0(u)} du \tag{4.112}$$

$$I_2 = \int_{u=0}^{u=1} G_2(u) e^{jk(f_0(u)+f_1(u))} du \tag{4.113}$$

The amplitude functions $G_1(u)$ and $G_2(u)$ are smoothly varying functions with respect to the exponential terms of the integrands. Then, the SPM for simple integrals can be applied to compute $I_1(u)$ and $I_2(u)$. The first integral can be expressed as follows:

$$I_1 = \sum_i I_{1s}^i + I_{10} + I_{11} \tag{4.114}$$

where the terms I_{1s}^i correspond to the stationary phase segments and the terms I_{10} and I_{11} to the boundary segments $u = 0$ and $u = 1$, respectively. The values of the stationary phase terms I_{1s}^i satisfy

$$f_{0u}(u_s^i) = 0 \tag{4.115}$$

and their contributions to the integral are given by

$$I_{1s}^i = G_1(u_s^i) e^{jkf_0(u_s^i)} \sqrt{\frac{2\pi j}{kf_0(u_s^i)}} \tag{4.116}$$

The contributions of the boundary segments are given by

$$I_{10} = stG_1(0) e^{jkf_0(0)} \sqrt{\frac{2}{k\,|f_{0uu}(0)|}} e^{-jsv^2(0)} F_s(v(0)) \tag{4.117}$$

$$I_{11} = stG_1(1) e^{jkf_0(1)} \sqrt{\frac{2}{k\,|f_{0uu}(1)|}} e^{-jsv^2(1)} F_s(v(1)) \tag{4.118}$$

where $v(x) = |f_{0u}(x)| \sqrt{\dfrac{k}{2\,|f_{0uu}(x)|}}$, $s = sign(f_{0uu})$, $t = sign(f_{0u})$ and $F_s(x)$ is the Fresnel function of the argument x.

The second integral can be expressed in a similar way:

$$I_2 = \sum_i I_{2s}^i + I_{20} + I_{21} \tag{4.119}$$

where the terms I^i_{2s} correspond to the stationary phase segments and the terms I_{20} and I_{21} correspond to the boundary segments $u = 0$ and $u = 1$, respectively. The values of the stationary phase terms I^i_{2s} satisfy

$$f_{0u}(u^i_s) + f_{1u}(u^i_s) = 0 \tag{4.120}$$

and their contributions to the integral are given by

$$I^i_{2s} = G_2(u^i_s)e^{jk\left(f_0(u^i_s)+f_0(u^i_s)\right)}\sqrt{\frac{2\pi j}{k\left(f_0(u^i_s)+f_0(u^i_s)\right)}} \tag{4.121}$$

The contributions of the boundary segments are given by

$$I_{20} = stG_2(0)e^{jk(f_0(0)+f_0(0))}\sqrt{\frac{2}{k\left|f_{0uu}(0)+f_{1uu}(0)\right|}}e^{-jsv^2(0)}F_s(v(0)) \tag{4.122}$$

$$I_{21} = stG_1(1)e^{jk(f_1(0)+f_0(1))}\sqrt{\frac{2}{k\left|f_{0uu}(1)+f_{1uu}(1)\right|}}e^{-jsv^2(1)}F_s(v(1)) \tag{4.123}$$

where $v(x) = |f_{0u}(x) + f_{1u}(x)|\sqrt{\dfrac{k}{2\left|f_{0uu}(x)+f_{1uu}(x)\right|}}$, $s = sign(f_{0uu}+f_{1uu})$ and $t = sign(f_{0u} + f_{1u})$.

The use of the SPM to integrate along the coordinate u requires that the contribution to the integral comes from a set of critical values: the stationary phase values u^i_s and the values at the limits of the integration interval, $u = 0$ and $u = 1$. As the surface is singly curved, these fixed values of the coordinate u are isoparametric segments contained upon it. Therefore, the contributions to the PO integral come from two kinds of segments on the patch: stationary phase segments and boundary segments (see Figures 4.8 and 4.9).

From a practical point of view, the only problem is to determine the stationary phase segments, that is, the values of u that verify conditions (4.115) and (4.120). Taking (4.32) into account, they can be written as:

$$\vec{k}_s \cdot \vec{r}'_{0u}(u) = 0 \tag{4.124}$$

$$\vec{k}_s \cdot (\vec{r}'_{0u}(u) + \vec{r}'_{1u}(u)) = 0 \tag{4.125}$$

On this kind of surface, the functions $\vec{r}_0'(u)$ and $\vec{r}_0'(u) + \vec{r}_1'(u)$ define the isoparametric curves $v = 0$ and $v = 1$, respectively (see Figures 4.10 and 4.11). Therefore, the vectors $\vec{r}_{0u}'(u)$ and $\vec{r}_{0u}'(u) + \vec{r}_{1u}'(u)$ define the tangent vectors of these curves. This means that (4.124) and (4.125) are satisfied for u values for which the wave vector is perpendicular to the vectors tangent to the curves of the surface formed by the boundaries $v = 0$ and $v = 1$.

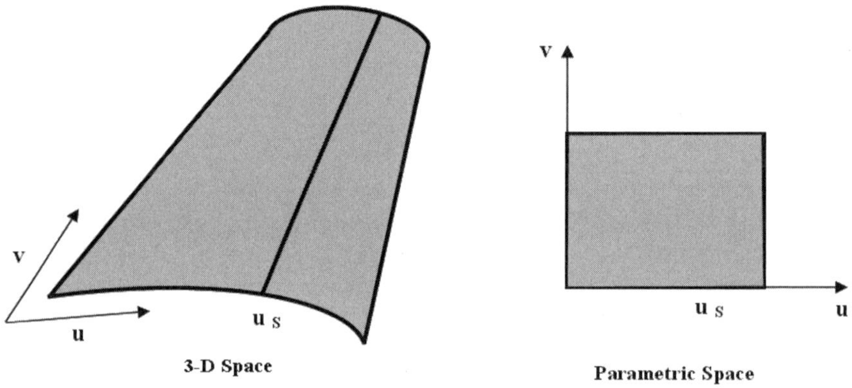

Figure 4.8 Stationary phase segments.

To determine if the contribution of a segment must be considered, it is necessary to determine if it is hidden. The main problem with the segments is that sometimes a segment can be partially illuminated and partially shadowed. As the integration method assumes that the contribution of the complete segment is considered, a criterion for the shadowing must be fixed. A typically used criterion is whether the middle point $v = 0.5$ is hidden and consider the complete segment to be illuminated or not, according to the illumination of the middle point.

Similar expressions can be obtained when the surface depends linearly on the parametric coordinate u.

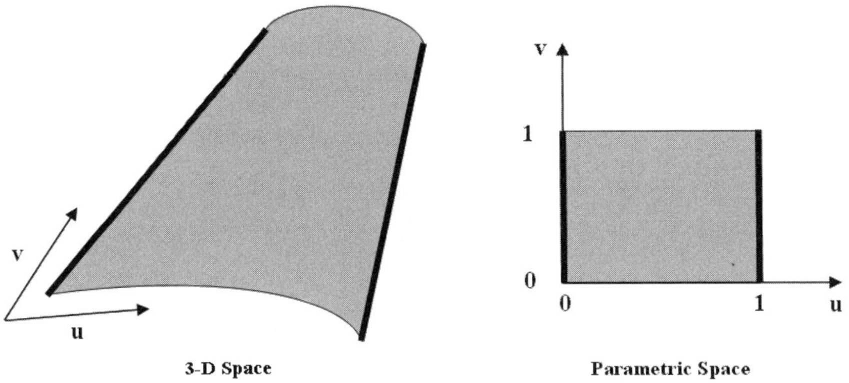

Figure 4.9 Boundary segments.

4.5 APPLICATION OF PO TO THE COMPUTATION OF HIGHER-ORDER REFLECTIONS

4.5.1 Multiple Reflections Between Plane Surfaces

Multiple reflections between plane surfaces can be computed very efficiently by combining geometrical optics with physical optics, resulting in a GO-PO [13] approach, which reduces the time needed to analyze multiple reflections compared to a complete PO-PO analysis.

The procedure for a double reflection will be presented first for a better understanding of the approach, and then it will be shown that it can easily be generalized to multiple order reflections. In a double reflection, GO is applied to study the reflection on the first facet, while Gordon's expression is used to evaluate the PO integral of the current induced in the second facet. It is assumed that the planar or spherical nature of the field incident on the facet is not changed after reflection. Then, the field reflected by the plane facet can be obtained by image theory [11]:

$$\vec{E}^{i'} = E_s^i \hat{s} - E_h^i \hat{h} \tag{4.126}$$

$$\vec{H}^{i'} = \frac{\vec{k}^{i'} \times \vec{E}^{i'}}{\eta} \tag{4.127}$$

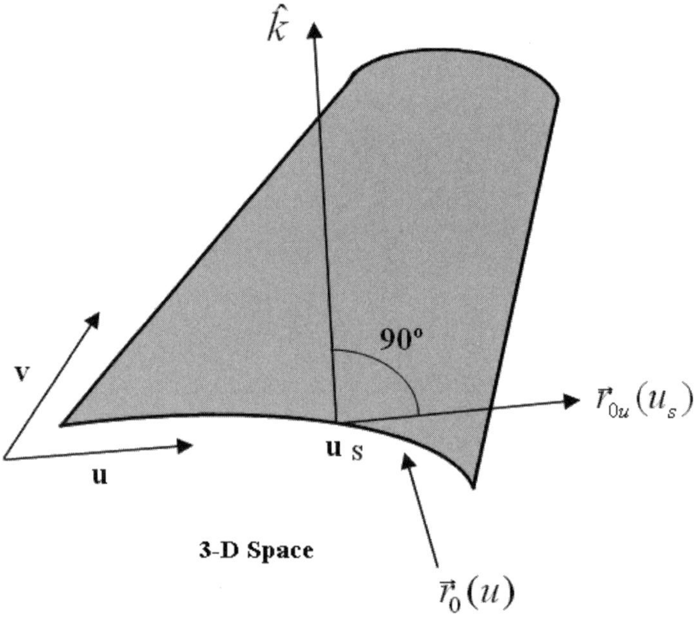

Figure 4.10 Stationary phase segment for the first integral.

where $E^i_{s,h}$ are the soft and hard components of the electric field of the plane wave incident on the planar facet, $\vec{E}^{i'}$ and $\vec{H}^{i'}$ are the images of the incident electric and magnetic fields, $\vec{k}^{i'}$ is the image of the direction of incidence, and $\eta = 120\pi$ is the impedance of the medium in free space.

The induced currents on the second planar surface can be computed with the PO approach:

$$\vec{J}'(\vec{r}') = 2\hat{n} \times \vec{H}^{i'} \tag{4.128}$$

As mentioned above, the electrical current in the second surface can be integrated using Gordon's method. It is important to take into account that the integral will be performed only in the illuminated part of the second surface (see Figure 4.12).

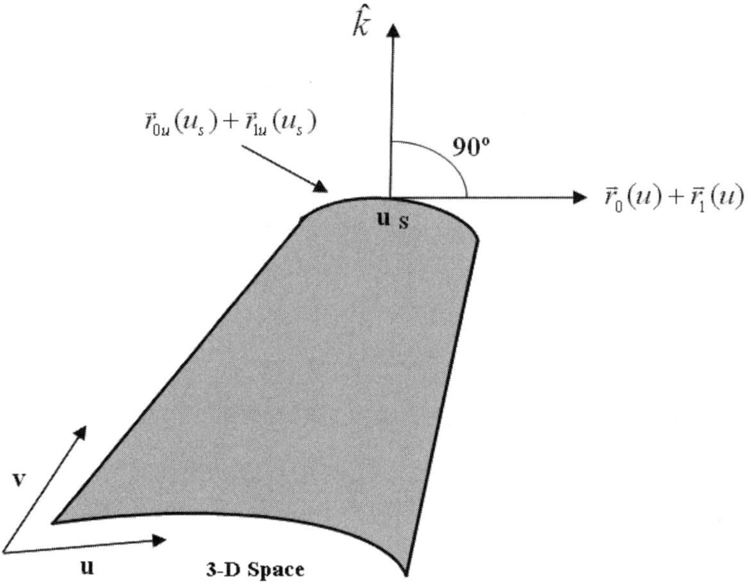

Figure 4.11 Stationary phase segment for the second integral.

In the case of an nth-order reflection, the procedure mentioned above can be easily generalized. The first $n-1$ reflections will be obtained using GO, and the nth reflection will be computed using Gordon's method, accounting for the illuminated part of the surface after the first $n-1$ reflections.

4.5.2 Double Reflection Between a Plane Surface and a Curved Surface

The problem can be classified into one of two cases, depending on whether the first surface is planar or curved. If the first surface is a plane, the reflected fields are computed using GO, following (4.126) and (4.127). After that, the current on the curved surface is computed using expression (4.128) and, as the surface is curved, Gordon's method cannot be applied. However, the current can be integrated by any of the other integration approaches described in Sections 4.3 and 4.4.

If the first surface is curved, the induced current in the curved surface is computed using the PO approach, given in expression (4.12). Then, the reflection

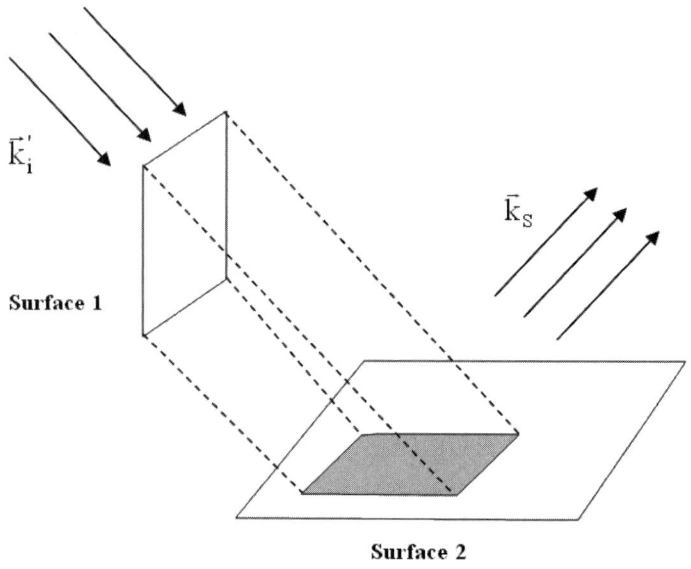

Figure 4.12 Illuminated part of the second surface in a double reflection between planar patches.

from the plane facet is treated according to the GO approach, and one obtains the image currents by applying image theory:

$$\vec{J}'(\vec{r}') = \left(J_\theta \hat{\theta} - J_\phi \hat{\phi}\right) e^{j\vec{k}^{i'} \cdot \vec{r}'} \tag{4.129}$$

where J_θ and J_ϕ are the θ and ϕ components of the induced current at the curved surface, respectively.

These image currents can be integrated by using any of the integration techniques described in Sections 4.3 and 4.4, considering only the illuminated part of the curved image surface [39].

4.5.3 Double Reflection Between Curved Surfaces Using the SPM

Double reflection between two curved surfaces is not easy to obtain from a computational point of view. However, an affordable approach can be developed using the SPM. To apply this approach and compute the double interactions between the patches, it is necessary to distinguish between the different kinds of critical points in the SPM.

4.5.3.1 Double Interaction Between Two Stationary Phase Points

These points present stationary behavior in phase for the two parametric coordinates (u_1, v_1, u_2, and v_2) of the two surfaces involved in the double interaction. The previous statement is translated, mathematically, to a system of four nonlinear equations in (4.130). The solution of the system gives the value of the two stationary points (u_{s1}, v_{s1}) and (u_{s1}, v_{s1}) involved in the analysis.

$$\left\{\begin{array}{l} f_{u1}(u_{s1}, v_{s1}, u_{s2}, v_{s2}) = 0 \\ f_{v1}(u_{s1}, v_{s1}, u_{s2}, v_{s2}) = 0 \\ f_{u2}(u_{s1}, v_{s1}, u_{s2}, v_{s2}) = 0 \\ f_{v1}(u_{s1}, v_{s1}, u_{s2}, v_{s2}) = 0 \end{array}\right. \Rightarrow \left\{\begin{array}{l} \left.\dfrac{\partial f}{\partial u_1}\right|_{(u_{s1}, v_{s1}, u_{s2}, v_{s2})} = 0 \\ \left.\dfrac{\partial f}{\partial v_1}\right|_{(u_{s1}, v_{s1}, u_{s2}, v_{s2})} = 0 \\ \left.\dfrac{\partial f}{\partial u_2}\right|_{(u_{s1}, v_{s1}, u_{s2}, v_{s2})} = 0 \\ \left.\dfrac{\partial f}{\partial v_2}\right|_{(u_{s1}, v_{s1}, u_{s2}, v_{s2})} = 0 \end{array}\right. \quad (4.130)$$

Once these points have been obtained, their contributions to the radiated field are very easy to compute iteratively. Two steps are considered. First, the radiated field at the second critical point (u_{s2}, v_{s2}) is obtained. This is the contribution of a simple stationary phase point (u_{s1}, v_{s1}) to an observation point located at the position given by the parametric coordinates (u_{s2}, v_{s2}), and can be obtained as explained in Section 4.4. After that, the induced current at the second critical point is obtained and integrated as indicated in that section. The main difficulty in calculating the second integral is that the second derivative of the phase with respect to u_2 and v_2 is needed. These derivatives cannot be obtained analytically and must be calculated numerically.

4.5.3.2 Double Interaction Between a Stationary Phase Point and a Boundary Point

In this case, one of the critical points is located on one of the boundaries that border the parametric space. Four boundaries are analyzed for each surface: $u_c = 0.0$, $u_c = 1.0$, $v_c = 0.0$, and $v_c = 1.0$. There are therefore eight possibilities: four in the boundary points belonging to the first surface, and four belonging to the second one. For each possibility, a parametric coordinate is fixed and the phase function must exhibit stationary behavior at the fixed coordinate.

In obtaining the radiated field, two cases can be distinguished, depending on the surface on which the boundary point is placed. In both cases the radiated field is obtained as described above, replacing the expression for a stationary phase point by the expression for a boundary point. As in the previous case, the derivative at the second surface must be evaluated numerically.

4.5.3.3 Double Interaction Between a Stationary Phase Point and a Vertex Point

In this case, the stationary phase point is easily obtained, as shown in Section 4.4 for the computation of single stationary phase points. The only difference is to replace the source position by the vertex position, in case the vertex-stationary phase point contribution is to be considered or replace the observer's position by the vertex if the contribution computed is the stationary phase point-vertex one.

The radiated field due to this double interaction is obtained in a similar way as the previous ones, using the expressions given in Section 4.4 for the contribution of a vertex.

4.6 PHYSICAL THEORY OF DIFFRACTION AND THE EQUIVALENT CURRENT METHOD

The treatment of edges is an important correction to the results obtained using PO for the analysis of radiation problems or scattering in complex structures. In the PO approach, the currents are considered to be associated with an infinite conductive plane. The scattered field is not considered to be a local phenomenon as in GO, but is obtained from the integration of the currents over the surface under study. Therefore, all the effects due to the fact that the surface is finite are not considered. Specifically, the real current is significantly different than the PO current in the border of the surface, sourcing a diffracted field that is not taken into account in the PO approach.

Ufimtsev [5] developed the PTD as a correction to the PO with this idea in mind. The hypothesis consists of assuming that the variation of the current with respect to the PO term is concentrated on the edge itself and the scattered field can therefore be obtained as the superposition of the PO field and the field radiated by the edge current.

The problem is to find the currents on the edge. The method proposed by Ufimtsev consists of subtracting the contribution of the PO approach from the field diffracted by an infinite edge. Therefore, the contribution due to the nonuniform current of the edge is isolated. The result is expressed as a function of the diffraction coefficients X and Y, related, obviously, with Keller's GTD:

$$X = \frac{\frac{1}{n}\sin\left[\frac{\pi}{n}\right]}{\cos\left[\frac{\pi}{n}\right] - \cos\left[\frac{\delta - \delta'}{n}\right]} \tag{4.131}$$

$$Y = \frac{\frac{1}{n}\sin\left[\frac{\pi}{n}\right]}{\cos\left[\frac{\pi}{n}\right] - \cos\left[\frac{\delta + \delta'}{n}\right]} \tag{4.132}$$

where $n\pi$ is the external angle of the wedge defined by the edge and δ and δ' represent the incidence and diffraction angles, respectively, measured in a plane perpendicular to the edge (see Figure 4.13).

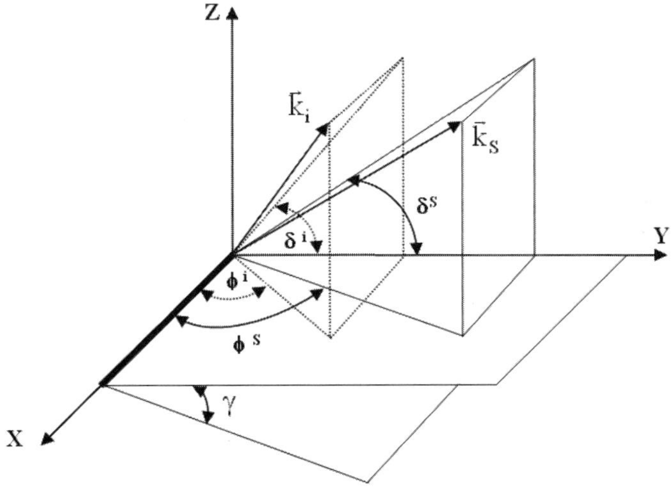

Figure 4.13 Angles associated with the PTD.

The diffraction coefficients of PTD, called f and g, take different values depending on which side of the wedge is illuminated (see Figure 4.14).

These coefficients are given outside of Keller's cone by:

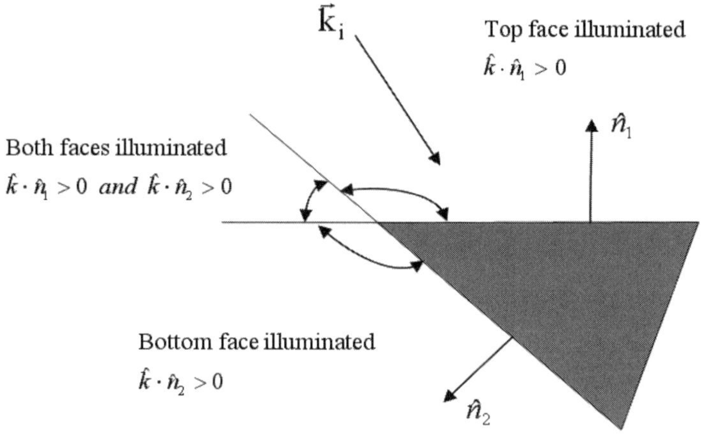

Figure 4.14 Different illuminated sides in the PTD.

- Bottom side illuminated, $\pi \leq \delta^i \leq \gamma$

$$f = X - Y - \left\{ \frac{1}{2} \tan\left[\frac{\delta^s - \delta^i}{2}\right] - \left[-\frac{1}{2} \tan\left[\gamma - \frac{\delta^s + \delta^i}{2}\right]\right]\right\} \tag{4.133}$$

$$g = X + Y - \left\{ \frac{1}{2} \tan\left[\frac{\delta^s - \delta^i}{2}\right] + \left[-\frac{1}{2} \tan\left[\gamma - \frac{\delta^s + \delta^i}{2}\right]\right]\right\} \tag{4.134}$$

- Top side illuminated, $0 \leq \delta^i \leq \gamma - \pi$

$$f = X - Y - \left\{ \frac{1}{2} \tan\left[\frac{\delta^s - \delta^i}{2}\right] - \left[-\frac{1}{2} \tan\left[\frac{\delta^s + \delta^i}{2}\right]\right]\right\} \tag{4.135}$$

$$g = X + Y - \left\{ \frac{1}{2} \tan\left[\frac{\delta^s - \delta^i}{2}\right] + \left[-\frac{1}{2} \tan\left[\frac{\delta^s + \delta^i}{2}\right]\right]\right\} \tag{4.136}$$

- Both sides illuminated, $\gamma - \pi \le \delta^i \le \pi$

$$\begin{aligned} f &= X - Y - \left\{ \frac{1}{2} \tan \left[\frac{\delta^s - \delta^i}{2} \right] - \left[-\frac{1}{2} \tan \left[\frac{\delta^s + \delta^i}{2} \right] \right] \right\} \\ &- \left\{ \frac{1}{2} \tan \left[\frac{\delta^s - \delta^i}{2} \right] - \left[-\frac{1}{2} \tan \left[\gamma - \frac{\delta^s + \delta^i}{2} \right] \right] \right\} \end{aligned} \tag{4.137}$$

$$\begin{aligned} g &= X + Y - \left\{ \frac{1}{2} \tan \left[\frac{\delta^s - \delta^i}{2} \right] - \left[-\frac{1}{2} \tan \left[\frac{\delta^s + \delta^i}{2} \right] \right] \right\} \\ &- \left\{ \frac{1}{2} \tan \left[\frac{\delta^s - \delta^i}{2} \right] + \left[-\frac{1}{2} \tan \left[\gamma - \frac{\delta^s + \delta^i}{2} \right] \right] \right\} \end{aligned} \tag{4.138}$$

The equivalent current method (ECM) is a way to calculate the diffraction from the edges [7]. The ECM was developed to compute the field at the caustics, where UTD predicts infinite fields, and also to calculate the field in regions outside Keller's cone. There are different approaches to these equivalent currents, derived from GTD or from PTD. In this case, the ECM is used together with the PTD coefficients mentioned above, as the field must be compatible with the PO approach.

The basic idea is to assume the existence of some electric and magnetic currents on the edge:

$$\vec{I}_e(\vec{r}') = \frac{2j\eta\hat{t} \cdot (\hat{t} \cdot \vec{E}^i)}{k \sin \beta^i \sin \beta^s} f(\delta^i, \delta^s) \tag{4.139}$$

$$\vec{I}_m(\vec{r}') = \frac{2j\eta\hat{t} \cdot (\hat{t} \cdot \vec{H}^i)}{k \sin \beta^i \sin \beta^s} g(\delta^i, \delta^s) \tag{4.140}$$

where $\hat{t}$ is the unit vector tangent to the edge, $\vec{E}^i$ and $\vec{H}^i$ are the incident electric and magnetic fields, respectively, and β^i and β^s are the angles between the edge and the incident and diffracted rays, respectively.

The electric and magnetic potential vectors can be obtained from these expressions at any point in space:

$$\vec{A}(\vec{r}) = \frac{\mu}{4\pi} \int_L \vec{I}_e(\vec{r}') \frac{e^{-jk|\vec{r}-\vec{r}'|}}{|\vec{r}-\vec{r}'|} dl' \tag{4.141}$$

$$\vec{F}(\vec{r}) = \frac{\epsilon}{4\pi} \int_L \vec{I}_m(\vec{r}') \frac{e^{-jk|\vec{r}-\vec{r}'|}}{|\vec{r}-\vec{r}'|} dl' \tag{4.142}$$

where the integral is solved along the edge (L).

The diffracted field of the edge can be immediately obtained as:

$$\vec{E}_\theta = (-j\omega A_\theta - j\omega\eta F_\phi) \cdot \hat{\theta} \tag{4.143}$$

$$\vec{E}_\phi = (-j\omega A_\phi + j\omega\eta F_\theta) \cdot \hat{\phi} \tag{4.144}$$

where A_θ , A_ϕ , F_θ, and F_ϕ are the spherical coordinates of the potential vectors.

If the edge is straight and the illumination is a plane wave (RCS computation), the scattered field can be computed analytically. In that case, the following expressions are obtained:

$$\vec{E}_\theta = -\frac{1}{2\pi}\left[-\frac{\hat{t}\cdot\vec{E}^i}{\sin^2\beta}f - \frac{\hat{t}\cdot\left(\vec{k}_i \times \vec{E}^i\right)}{\sin^2\beta}g\right] Le^{jk_tL}SINC\left[\frac{k_tL}{\pi}\right]\frac{e^{-jkr}}{r}\hat{\theta} \tag{4.145}$$

$$\vec{E}_\phi = -\frac{1}{2\pi}\left[-\frac{\hat{t}\cdot\vec{E}^i}{\sin^2\beta}f + \frac{\hat{t}\cdot\left(\vec{k}_i \times \vec{E}^i\right)}{\sin^2\beta}g\right] Le^{jk_tL}SINC\left[\frac{k_tL}{\pi}\right]\frac{e^{-jkr}}{r}\hat{\phi} \tag{4.146}$$

where k_t is the component of the wave vector tangential to the edge.

In other cases, the integrals of (4.141) and (4.142) can be calculated numerically or by using the SPM for simple integrals.

The reasoning given above is purely mathematical and has no physical foundation. In fact, the equivalent currents obtained depend on the point of observation of the radiated field, which lacks physical sense. However, it is a proper computation tool. For an effective use of this approach, it is necessary to assume that the equivalent currents obtained for an infinite edge can be assigned to a large enough edge and perform a further integration over the finite length of the edge.

4.7 INTRODUCTION OF RADAR ABSORBING MATERIALS (RAM) IN THE PO FORMULATION

If the body is not PEC, the formulation for the PO currents changes. The effect of the material on the electric current can be considered by using the Fresnel reflection coefficients. Moreover, the magnetic current is not zero, but depends on the reflection coefficients of the material. The formulation for these currents is given in [15]:

$$\vec{J}_S(\vec{r}') = \hat{n} \times \begin{bmatrix} 1-\Gamma_s & 0 \\ 0 & 1-\Gamma_h \end{bmatrix} \begin{bmatrix} \vec{H}^i_h(\vec{r}') \\ \vec{H}^i_h(\vec{r}') \end{bmatrix} \tag{4.147}$$

$$\vec{M}_S(\vec{r}') = -\hat{n} \times \begin{bmatrix} 1+\Gamma_s & 0 \\ 0 & 1+\Gamma_h \end{bmatrix} \begin{bmatrix} \vec{E}^i_h(\vec{r}') \\ \vec{E}^i_h(\vec{r}') \end{bmatrix} \tag{4.148}$$

where $\hat{n}$ is the unit normal vector of the surface at the point $\vec{r}'$, Γ_s and Γ_h are the Fresnel reflection coefficients, and $\vec{H}^i_{s,h}$ and $\vec{E}^i_{s,h}$ are the soft and hard components of the incident magnetic and electric field at the point on the surface, respectively. The soft component of the incident field is the component of the electric field parallel to the plane of incidence while the hard component represents the component of the electric field perpendicular to the plane of incidence [40].

The total scattered field can be computed as:

$$\vec{E}^s = \vec{E}^s_J + \vec{E}^s_M \tag{4.149}$$

where $\vec{E}^s_J$ and $\vec{E}^s_M$ are the scattered fields due to the electric ($\vec{J}$) and the magnetic current ($\vec{J}$), respectively:

$$\vec{E}^s_J = j\frac{\eta}{2\lambda}\frac{e^{-jkr}}{r}\hat{k}_s \times \left(\hat{k}_s \times \int_S \vec{J}_S(\vec{r}')e^{j\vec{k}_s\cdot\vec{r}'}\,dS'\right) \tag{4.150}$$

$$\vec{E}^s_M = j\frac{1}{2\lambda}\frac{e^{-jkr}}{r}\left(\hat{k}_s \times \int_S \vec{M}_S(\vec{r}')e^{j\vec{k}_s\cdot\vec{r}'}\,dS'\right) \tag{4.151}$$

The integrals of (4.150) and (4.151) can be calculated with any of the procedures presented in this chapter.

References

[1] C. Uluisik, G. Cakir, M. Cakir, and L. Sevgi. "Radar cross section (RCS) modeling and simulation, part 1: A tutorial review of definitions, strategies and canonical examples." *IEEE Antennas and Propagation Magazine*, 50(1):115–126, February 2008.

[2] L. B. Felsen and N. Marcuvitz. *Radiation and Scattering of Waves*. Prentice-Hall, 1973.

[3] G. T. Ruck, D. E. Barrick, W. D. Stuart, and C. K. Krichbaum. *Radar Cross Section Handbook. Vol. 1*. Plenum Press, 1985.

[4] E. F. Knott, J. F. Shaeffer, and M. T. Tuley. *Radar Cross Section. Its Prediction, Measurements, and Reduction*. Artech House, 1985.

[5] P. Y. Ufimtsev. "Elementary edge waves and the physical theory of difracction." *Electromagnetics*, 11(2):125–160, April-June 1991.

[6] J. M. Rius, M. Ferrando, and L. Jofre. "High-frequency RCS of complex radar targets in real-time." *IEEE Transactions on Antennas and Propagation*, 41(9):1308–1319, September 1993.

[7] C. E. Ryan and L. Peters. "Evaluation of edge-diffracted fields including equivalent currents for caustic regions." *IEEE Transactions on Antennas and Propagation*, AP-17(3):292–299, May 1969.

[8] V. H. Weston. "Theory of absorbers in scattering." *IEEE Transactions on Antennas and Propagation*, 11(5):578–584, September 1963.

[9] D. M. Elking, J. M. Roedder, D. D. Car, and S. D. Alspach. "A review of high-frequency radar cross section analysis capabilities at McDonnell Douglas aerospace." *IEEE Antennas and Propagation Magazine*, 37(5):33–42, October 1995.

[10] D. Andersh, J. Moore, S. Kosanovich, D. Kapp, R. Bhalla, R. Kipp, T. Courtney, A. Nolan, F. German, J. Cook, and J. Hughes. "Xpatch 4: The next generation in high frequency electromagnetic modeling and simulation software." *Radar Conference*, pp. 844–849, 2000.

[11] M. F. Cátedra and J. Pérez-Arriaga. *Cell Planning for Wireless Communications*. Artech House, 1999.

[12] H. Kobayashi, Y. Tokumitsu, and K. Hongo. "RCS calculation software system using PTDTC." *International Conference on Computational Electromagnetics and Its Applications. (ICCEA'99)*, pp. 569–572, 1999.

[13] L. Lozano, M. J. Algar, M. Blanco, I. González, and M. F. Cátedra. "Computer code for the efficient calculation of the RCS of complex target considering multiple bounces." *IEEE Antennas and Propagation International Symposium*, 2008.

[14] M. Domingo, F. Rivas, J. Pérez, R. P. Torres, and M. F. Cátedra. "Computation of the RCS of complex bodies modeled using NURBS surfaces." *IEEE Antennas and Propagation Magazine*, 37(6):36–47, December 1995.

[15] F. Saez de Adana, I. González, O. Gutiérrez, P. Lozano, and M. F. Cátedra. "Method based on physical optics for the computation on the radar cross section including diffraction and double effects of metallic and absorbing bodies modeled by parametric surfaces." *IEEE Transactions on Antennas and Propagation*, 52(12):3295–3303, December 2004.

[16] W. B. Gordon. "Far-field approximation to the Kirchoff-Helmholtz representations of scattered field." *IEEE Transactions on Antennas and Propagation*, AP-23(4):590–592, July 1975.

[17] D. S. Jones and M. Kline. "Asymptotic expansion of multiple integrals and the method of stationary phase." *J. Math. Phys.*, 37:1–28, 1958.

[18] J. C. Cooke. "Stationary phase in two dimensions." *IMA Journal of Applied Mathematics*, 29:25–37, 1982.

[19] A. Boag. "A fast physical optics (FPO) algorithm for high frequency scattering." *URSI Radio Science Meeting*, 2000.

[20] A. Boag and E. Michielssen. "A fast physical optics (FPO) algorithm for double-bounce scattering." *IEEE Transactions on Antennas and Propagation*, 52(1):205–212, January 2004.

[21] M. Cheng, M. Wang, Y. Zhang, and C. Liang. "Parallel NURBS PO method based on the Ludwig algorithm and its application to the RCS computation." *Microwaves Conference Proceedings, Vol. 4*, 2005.

[22] C. A. Siller, Jr. "Evaluation of the radiation integral in terms of end-point contributions." *IEEE Transactions on Antennas and Propagation*, 23(5):743–745, September 1975.

[23] H. Ikuno. "Calculation of far-scattered fields by the method of stationary phase." *IEEE Transactions on Antennas and Propagation*, 27(2):199–202, March 1979.

[24] H. Nakano, S. R. Kerner, and N. G. Alexopoulos. "The moment method solution for printed wire antennas of arbitrary configuration." *IEEE Transactions on Antennas and Propagation*, 36(12):1667–1674, December 1988.

[25] J. A. Shifflet. "CADDRAD: A physical optics radar/radome analysis code for arbitrary 3D geometries." *IEEE Antennas and Propagation Magazine*, 39(6):73–79, December 1997.

[26] O. M. Conde, J. Pérez, and M. F. Cátedra. "Stationary phase method application for the analysis of radiation of complex 3-D conducting structures." *IEEE Transactions on Antennas and Propagation*, 49(5):724–731, May 2001.

[27] O. M. Conde, F. Saez de Adana, and M. F. Cátedra. "A comparison between two high-frequency techniques applied to the analysis of on-board antennas." *Microwave and Optical Technology Letters*, 36(5):415–417, March 2003.

[28] J. A. Stratton. *Electromagnetic Theory*. McGraw-Hill, 1941.

[29] R. Mittra. *Computer Techniques for Electromagnetics*. Pergamon Press, 1973.

[30] A. L. Maffet. *Topics for a Statistical Description of Radar Cross Section*. John Wiley and Sons, 1989.

[31] E. P. Knott and T. B. A. Senior. "Comparison of three high-frequency diffraction techniques." *Proceedings of the IEEE*, 62(11):1468–1474, November 1974.

[32] J. D. Faires and R. L. Burden. *Numerical Analysis*. Brooks Cole, 2002.

[33] A. C. Ludwig. "Computation of radiation patterns involving numerical double integration." *IEEE Transactions on Antennas and Propagation*, 16(6):767–769, November 1968.

[34] N. N. Youssef. "Radar cross section of complex targets." *Proceedings of the IEEE*, 77(5):722–734, May 1989.

[35] M. Domingo, R. P. Torres, and M. F. Cátedra. "Contribution of double interaction between edges and facets to monostatic RCS of complex targets." *IEEE Antennas and Propagation International Symposium, Vol. 2*, pp. 937–940, 1992.

[36] D. Klement, J. Peissner, and V. Stein. "Special problems in applying the physical optics method for backscattering computation of complicated objects." *IEEE Transactions on Antennas and Propagation*, 36(2):228–238, February 1988.

[37] Z. Altman and R. Mittra. "A technique for extrapolating numerically rigorous solutions of electromagnetic scattering problems to higher frequencies and their scaling properties." *IEEE Transactions on Antennas and Propagation*, 47(4):744–751, April 1999.

[38] M. F. Cátedra, C. Delgado, S. Luceri, and F. Saez de Adana. "Efficient procedure for computing fields created by current modes." *Electronic Letters*, 39(10):763–765, May 2003.

[39] F. Saez de Adana, S. Nieves, E. García, I. González, O. Gutiérrez, and M. F. Cátedra. "Calculation of the RCS from the double reflection between plane facets and curved surfaces." *IEEE Transactions on Antennas and Propagation*, 51(9):2509–2512, September 2003.

[40] C. A. Balanis. *Advanced Engineering Electromagnetics*. John Wiley and Sons, 1989.

Chapter 5

Acceleration Techniques

5.1 ACCELERATION TECHNIQUES APPLIED TO HIGH-FREQUENCY PROBLEMS

The application of high-frequency techniques to complex problem analysis modeled by parametric surfaces has several advantages. The main advantage is the accuracy in the representation of the body using a low amount of information. However, the computational treatment associated with the application of these techniques to parametric surfaces is higher than when using simpler models. For example, the ray-tracing in a GO/UTD problem is computationally more expensive for a model composed by parametric surfaces than for a model composed by plane facets. In the first model, the shadowing test and the flash-point search (searching for the reflection and diffraction points) is performed by the CGM through minimization. In the plane facets models, analytical expressions may be used for both types of tasks. Something similar can be stated about the determination of the illuminated surfaces and the computation of the PO integral in the application of that technique to complex problems, especially for the RCS computation. The contribution of a plane facet to the RCS can be obtained by using Gordon's method, which is completely analytical. However, as discussed in Chapter 4, the integration techniques are more complicated when parametric surfaces are involved.

In complex problems, the analysis is usually performed for multiple directions of the space both in on-board antennas and in RCS computation problems. Therefore, an analysis that uses brute force (analyzing every surface of the model and its contribution to every direction of the space) in a complex model becomes practically unaffordable from a computational point of view. The number of surfaces under consideration when one direction of the space is treated must be limited.

Performing a previous process allows selecting the surfaces that may potentially contribute in a given direction, in the shadowing of the ray, in the searching of a flash point, or in the computation of the PO integral. Use of the previous process considerably reduces the number of surfaces analyzed for each direction and, in turn, the computational cost of the problem.

Techniques known as ray-tracing acceleration are used to reduce the number of surfaces and the cost. The name of ray-tracing comes because these techniques have been widely used in the world of CAGD to accelerate the process of illuminating a scene with many objects [1, 2]. This process poses a ray-tracing problem, and its adaptation to the GO/UTD problems, which are based on ray-tracing, is straightforward. This chapter shows that these techniques may also be applied to the RCS computation using PO. RCS is not a ray-tracing problem; however, at high frequencies, it may have some common features with ray-tracing, and therefore, RCS computation problems may use similar acceleration techniques.

Ray-tracing acceleration techniques have been widely applied to the analysis of mobile communication propagation problems when models are created with plane facets [3–8]. The most classical techniques are the binary space partitioning (BSP) [9], the space volumetric partitioning (SVP) [1, 2, 10], and the angular Z-buffer [3, 10]. In BSP, a tree with the facets of the space is created to analyze the facets in the branches where the source and the observer are placed. In the SVP the space is divided into cubic regions that follow Cartesian coordinates. The AZB [3, 10] has space divided into solid angular regions according to spherical coordinates that use the source as a reference. As can be seen, to reduce the number of facets involved in the shadowing problem for a given direction of the space, these techniques divide the space into different regions and classify the facets of the model in the different regions. These techniques have been improved to introduce hierarchically partitioned structures [11] or the quadtree [12] and octree techniques [13], which are based on the recursive division of the space to accelerate the shadowing process.

The reduction of the number of facets under consideration when the shadowing of a ray must be determined receives most of the effort. However, some efforts have been performed to reduce the computational time necessary to determine the reflections in the facets of the model as the algorithm based on the double-tree image theory [14]. In the double-tree image theory, a tree of images is used to obtain the multiple reflections that are possible between the source and the observer. Another algorithm with this objective is the 3-D tetrahedron algorithm [15]. In the 3-D tetrahedron algorithm, space is divided into tetrahedral regions that depend on the number of model vertices. Both algorithms make use of the possibility of using the image theory [16] to compute the reflection in a plane facets model.

All of these techniques have been applied to the analysis of propagation in mobile communications for plane facets models by using GO/UTD. Although some efforts have been dedicated to on-board antennas problems [17], bodies that are mostly composed of plane facets, such as satellites and ships, were considered. In these cases, two features are considered when using a ray-tracing acceleration technique: image theory can be used to compute the reflection, and the shadowing problem is computationally more expensive than flash-point searching. This is because the possibility of performing both tasks analytically increases the number of intersection tests necessary for a brute force analysis more than the number of interrogations needed for a flash point search, as was proved in [3, 10].

If the body is modeled by parametric surfaces, the approach must differ. The shadowing problem has a high computational cost, but the flash point search is also expensive. When double- and higher-order effects are involved, the necessary minimization is computationally more expensive, not only due to possible combinations of model surfaces in which the multiple effect can occur, but also for the number of function variables of the function to minimize. For example, the determination of a double reflection requires the minimization of a four-variable function, whereas a simple reflection or the determination of an intersection ray surface requires a two-variable function. The computational cost associated with the CGM increases exponentially with the number of variables. Therefore, it is important to reduce the number of minimizations performed to determine the presence of a double reflection. The resolution of the PO integral in a RCS computation is similar. Solving the PO integral over a parametric surface is a computationally expensive task. To develop a method that handles complex bodies in an affordable amount of time, it is necessary to reduce the number of PO integrals to evaluate for a given direction.

This chapter presents the application of acceleration techniques to the resolution of high-frequency problems over bodies modeled by parametric surfaces. The techniques considered are based on the AZB technique, but they are adapted to the shadowing problem and are applied to arbitrarily-shaped surfaces as well as for the flash point search over these kinds of surfaces. The AZB is combined with the SVP when the analysis is performed in the near zone. These two techniques are introduced and reviewed in this chapter in the context of their problem-solving application. Ray-tracing and RCS computation problems are in the following sections. Before applying the AZB and the SVP, some simple criteria can be applied to reduce of the number of surfaces involved in the analysis and the memory required in both techniques. The simple criteria are also covered in this chapter.

5.2 ELIMINATION OF SURFACES USING SIMPLE CRITERIA

There are two simple criteria that can be applied to the surfaces of the model when the flash point search or shadowing tests are performed. In some situations, these criteria can eliminate almost 50% of the model surfaces. They are the normal vector criterion (backface culling [10]) and intersection with the bounding box.

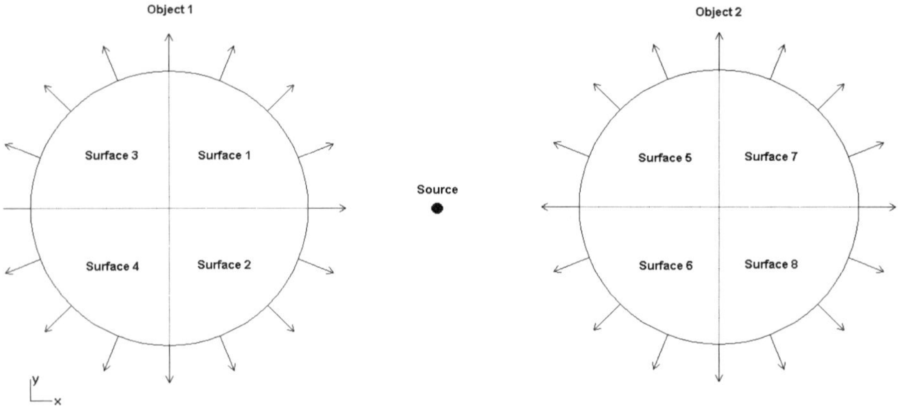

Figure 5.1 An example of surface visibility.

5.2.1 Normal Vector Criterion

The normal vector criterion determines the illuminated surfaces of the model to select the surfaces that can be either shadow a ray in a UTD analysis or the surfaces contributing to the PO integral. The illuminated surfaces of the model are those that satisfy the following condition:

$$\hat{k}_i \cdot \hat{n} < 0 \tag{5.1}$$

where $\hat{k}_i$ is the direction of incidence and $\hat{n}$ is the normal vector to the surface. The surfaces that accomplish the previous criterion are considered visible, and the rest are considered shadowed by illumination. The visible surfaces are determined through the consideration of sample points over the surface. The criterion (5.1) is applied to each sample point considering its normal vector. If one of the points satisfies the condition (5.1), the surface will be visible. Figure 5.1 shows an example with a source in the middle of two cylinders that are each formed by four surfaces. Figure 5.1 represents the sample points at each surface. Only surfaces 1, 2, 5, and 6 are illuminated from the source.

This criterion is usually used to determine the surfaces that can shadow a ray in UTD. However, it may be used to search for the reflection of diffraction points over the model [10]. The visible surfaces can only contain one of these points.

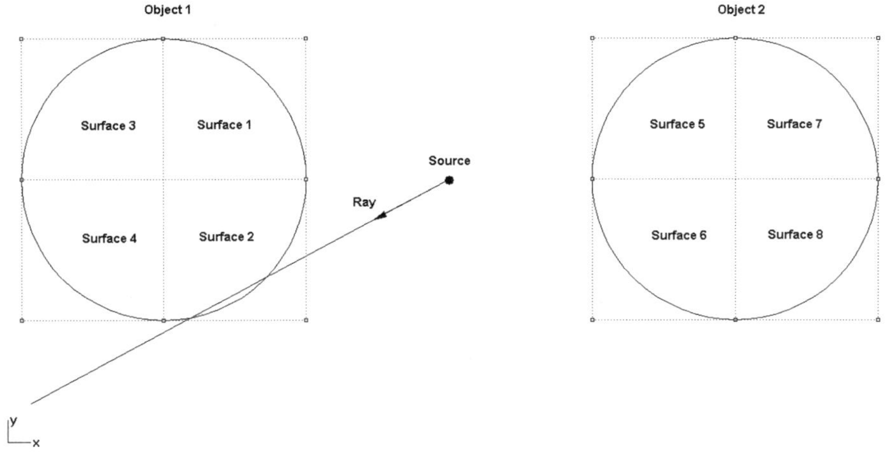

Figure 5.2 An example of a cut with the bounding box.

5.2.2 Intersection with the Bounding Box

The intersection with the bounding box criterion is used to discard surfaces when the shadowing of a ray is determined. The shadowing test based on the minimization presented in Chapter 3 is applied only if a ray intersects with the bounding box to the surface. This test is avoided if the ray does not intersect with the box. Figure 5.2 shows an example with the example in Figure 5.1. Only four of the eight surfaces of Figure 5.1 are illuminated, and only the bounding box to surface 2 intersects the ray. Therefore, the minimization shadowing test must be applied to one model surface and the computation time necessary for the shadowing test is reduced.

5.3 CLASSICAL TECHNIQUES FOR THE RAY-TRACING ACCELERATION FOR THE SHADOWING TEST

Most of the ray-tracing acceleration tests reduced the computation time necessary for the shadowing test of plane facets models. In this chapter, two methods (the AZB or the SVP) are generalized for parametric surfaces as well as for the determination of the flash points and the RCS computation. This section presents the basis of these two techniques as a previous step necessary for a better comprehension of their application to complex models that are composed of parametric surfaces.

5.3.1 The Angular Z-Buffer (AZB) Algorithm

The AZB technique was first applied to the analysis of outdoor propagation in models composed by plane facets [3], but it can be generalized to parametric surfaces. It is based on the light buffer technique [18] used for computer graphic design.

In the AZB, the space is divided into spherical sectors called anxels (angular elements) from the source point. These elements are defined by the spherical coordinates theta (θ) and phi (ϕ) of the fixed coordinate system located on the source (see Figure 5.3).

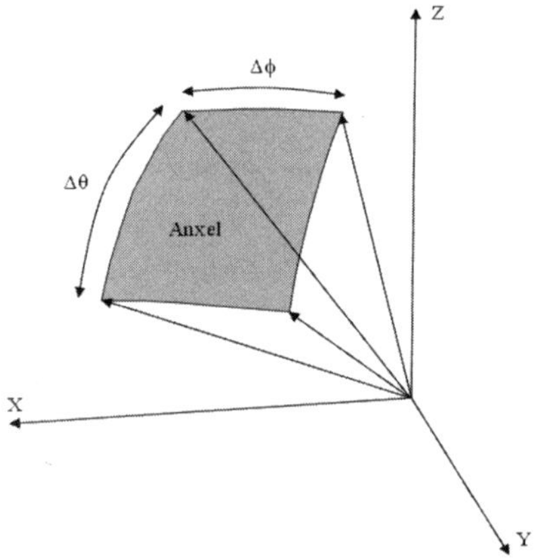

Figure 5.3 Definition of the anxel.

The anxel where each facet belongs is computed from the spherical coordinates of its vertices. Figure 5.4 represents the anxels in the (θ, ϕ) plane (AZB plane), with an example of the surfaces that lie in each anxel.

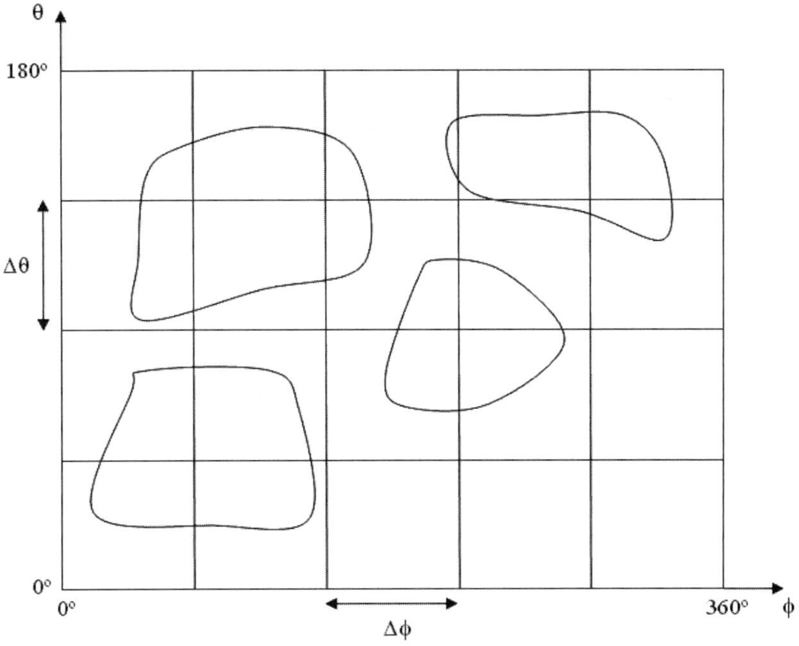

Figure 5.4 An example of surface storage in the AZB plane.

This information is stored in the matrix referred to as the AZB. It depends on the source point and the environmental model exclusively. For each anxel, the facets are arranged according to their distance to S. Figure 5.5 shows a simple 2-D outdoor scene with a space partitioning in 8 anxels. Table 5.1 shows the storage of the surfaces in the AZB matrix.

The spherical coordinates locate the anxel that corresponds to a given observation point (O) and this information is used for the shadowing test. Only the surfaces located in the corresponding anxel can shadow the ray. Considerably fewer intersection tests are needed to determine if the ray is shadowed. The reduction in the number of tested facets depends on the values of $\Delta\theta$ and $\Delta\phi$; the number of facets by anxel decreases with the number of anxels. On the other hand, the number of regions is limited by the available memory.

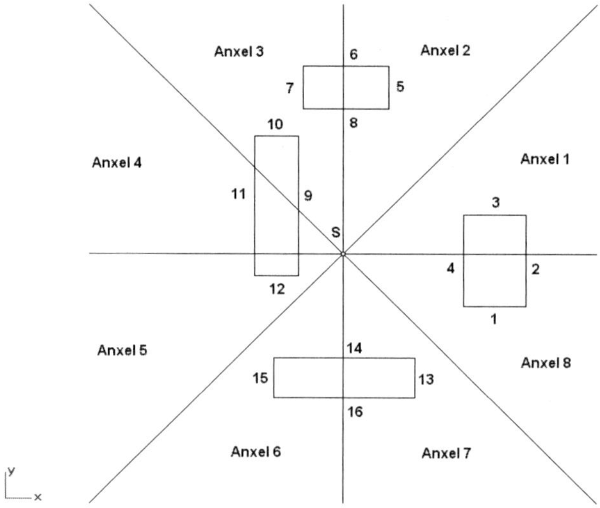

Figure 5.5 An example of a 2-D scene and its division into anxels.

Table 5.1
The Distribution of the Model Surfaces in the Corresponding Anxels

Anxel	*Surfaces*
1	2, 3, 4
2	5, 6, 8
3	6, 7, 8, 9, 10
4	9, 11
5	9, 11, 12
6	14, 15, 16
7	13, 14, 16
8	1, 2, 4

5.3.2 Space Volumetric Partitioning (SVP)

The space volumetric partitioning (SVP) technique involves the division of space into volumes called voxels (volume elements). The voxels are parallelepipeds. A 2-D example is shown in Figure 5.6. The surfaces that belong to each voxel are obtained after the space is divided. These surfaces are those which are totally or partially inside the volume. This information is loaded into the SVP matrix. The SVP matrix of Figure 5.6 is shown in Table 5.2.

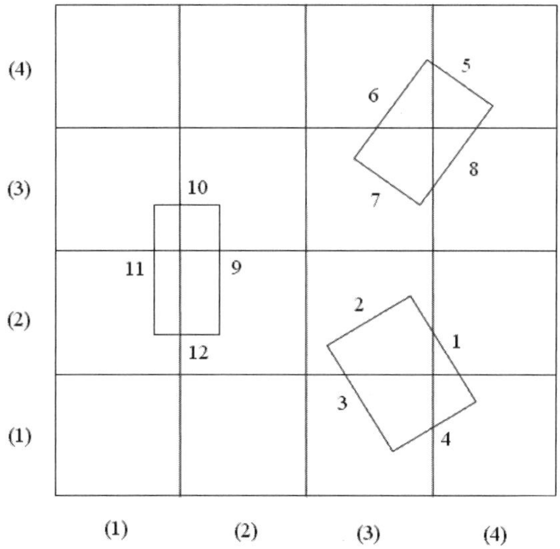

Figure 5.6 An example of a 2-D scene and its division into voxels.

When a shadowing test for a ray path source-observer is completed, the voxel where the source lies is determined. All of the surfaces loaded in the SVP matrix for this voxel are interrogated. If none of the surfaces shadows the ray, the surfaces contained in the next voxel that the ray crosses are considered for the intersection test. The procedure is repeated until a facet of the model shadows the ray or until the observation point is reached (in this case the ray is not shadowed and contributes to the total field). The SVP algorithm is applied in a similar fashion and independently of the effect considered. The source can be the transmitter antenna or a reflection or diffraction point. The observer can be an observation point or a reflection or diffraction point. Therefore, the division in voxels is the same for all of the effects. The SVP matrix only depends on the geometrical model and not on the source position and the effect considered.

Table 5.2

The Distribution of the Model Surfaces in the Corresponding Voxels

Voxel	*Surfaces*
(1,1)	
(1,2)	
(1,3)	3, 4
(1,4)	1, 4
(2,1)	11, 12
(2,2)	9, 12
(2,3)	2, 3
(2,4)	1
(3,1)	10, 11
(3,2)	9, 10
(3,3)	6, 7, 8
(3,4)	8
(4,1)	
(4,2)	
(4,3)	5, 6
(4,4)	5, 8

5.4 APPLICATION OF THE AZB TO BODIES MODELED BY PARAMETRIC SURFACES

5.4.1 Application to the Shadowing of the Direct Ray

Following the procedure described in Section 5.3.1, surfaces that are visible from the source are stored in the corresponding anxels of the AZB plane. This information will be used to reduce the number of intersection tests once the AZB matrix is created.

It is important to consider all the directions in which a given arbitrarily shaped surface can shadow a ray. The arbitrary shape of the surface makes obtaining the anxels where all of the surface points are included very difficult. Therefore, a good alternative is to consider a simpler shape as the bounding box. The anxels that include the bounding box can be obtained with its eight vertices easily. Therefore, these anxels are obtained with a low computational cost. As a drawback, the volume of the box is greater than the surface. Some surfaces can be included in anxels in which the surface is not contained.

In order to improve the features of the AZB, only the surfaces that satisfy the backface culling criterion are included in the AZB matrix. Moreover, the surfaces are stored following the format known as the painter's algorithm [19]. The surfaces are stored according to their distance from the source. Therefore, during the shadowing test, the surfaces closest to the source will be considered first because these are the most probable surfaces that can shadow the ray.

The procedure to build the AZB matrix is as follows:

1. The eight vertices of the bounding box of surface i are obtained. The first visible surface is i.
2. The spherical coordinates (θ, ϕ) of each vertex taking the source as a reference are obtained. Figure 5.7 shows an example.

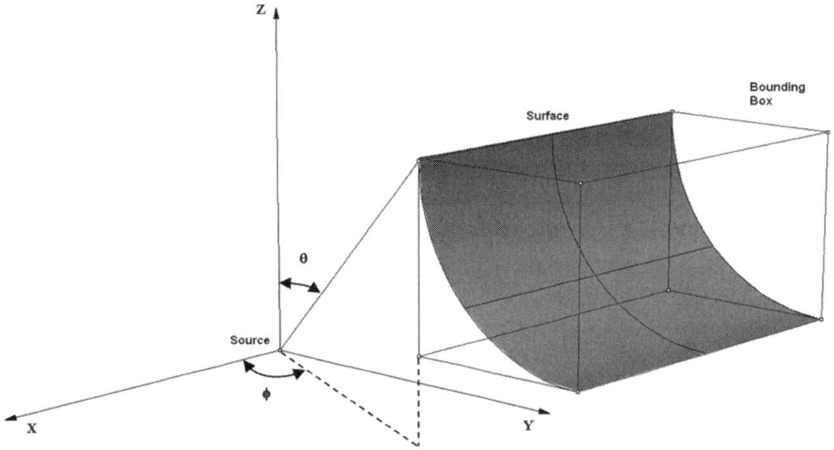

Figure 5.7 Spherical coordinates for a vertex in the box.

3. The maximum and minimum values of all the vertices of the previous step are obtained. These values are defined as θ_{max}, θ_{min}, ϕ_{max}, ϕ_{min}. The margins,

$\Delta\theta_s$ and $\Delta\phi_s$, occupied by the surface can be determined. The anxels in which surface i must be stored can be obtained from these margins. Figure 5.8 shows this situation.

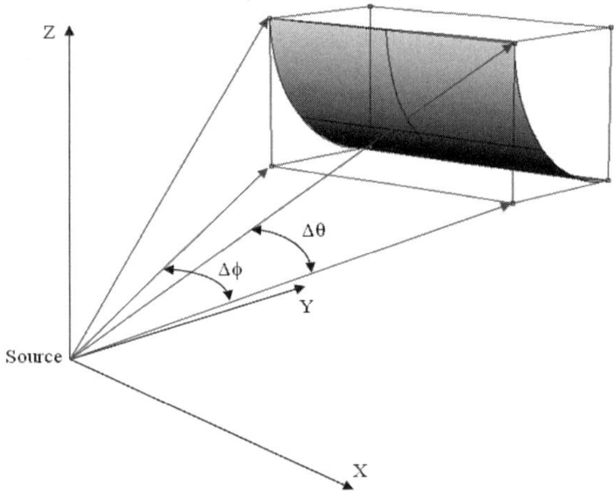

Figure 5.8 Angular margins of the surface.

It is necessary to check if the maximum and minimum values correspond to the real ones because there are special cases in which these values are obtained for points that do not correspond to the box vertices. It is necessary to correct these values to avoid mistakes. The following condition indicates when it is necessary to perform the correction:

$$\Delta\phi_s > \pi \tag{5.2}$$

Figure 5.9 shows the two situations that may appear for the example known as the special case in ϕ. These cases appear when the surface is on the X axis of the reference system defined by the source because the ϕ coordinates go from 360° to 0° in this axis. In this case, coordinates ϕ_{max} and ϕ_{min} must be modified with the values between two consecutive vertices that have an angular difference greater than 180° between them. If the situation of Figure 5.9(a) occurs and the vertices are ordered in ascending order according the coordinate, the difference between ϕ_2 and ϕ_3 is greater than 180° and it must be updated $\phi_{max} = \phi_2$ and $\phi_{min} = \phi_3$. The other special case occurs when the surface is included in all

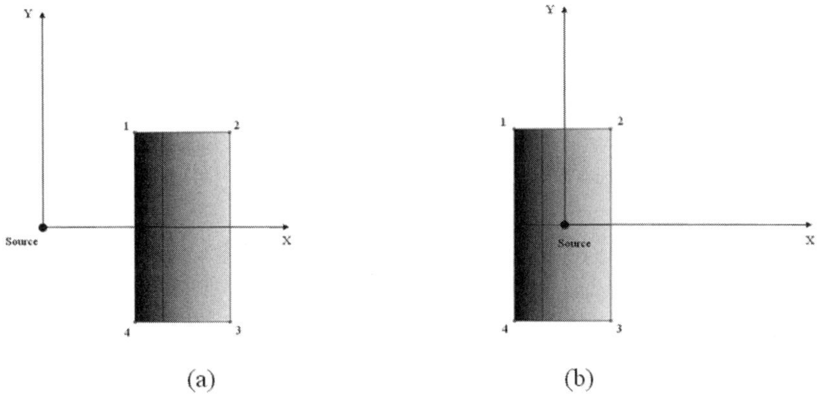

Figure 5.9 Special cases in ϕ.

of the ϕ ($0° - 360°$) regions. This situation appears when the source is above or below the surface. Figure 5.9(b) illustrates this situation. In this case, it is necessary to correct the values of the coordinates θ_{max} and θ_{min} to $180°$ and $0°$, respectively. Figure 5.10 indicates that the resulting AZB matrix depends on the case.

4. The next step is to determine the anxels in which surface i will be stored. The placement of i depends on the minimum and maximum values of the spherical coordinates and on the special case. A rectangle in the AZB plane is created with the calculated values and the corresponding cells of the matrix will be filled:

   - If there is not a special case, the surface i must be stored in all the anxels that contain, at least partially, the rectangle formed by the maximum and minimum coordinates. In the example of Figure 5.11(a), surface i is included in the cells (2,2), (2,3), (2,4), (3,2), (3,3), (3,4), (4,2), (4,3), and (4,4), marked in gray.
   - If the surface is a special case in ϕ, the anxels corresponds with a situation similar to the one shown in Figure 5.11(b). The external cells to the rectangle defined by the maximum and minimum coordinates must be considered. These are those that contain the direction which accomplish $\phi < \phi_{min}$ or $\phi < \phi_{max}$. In the example of Figure 5.11(b), those regions are (2,1), (2,2), (2,4), (2,5), (3,1), (3,2), (3,4), (3,5), (4,1), (4,2), (4,4), and (4,5).

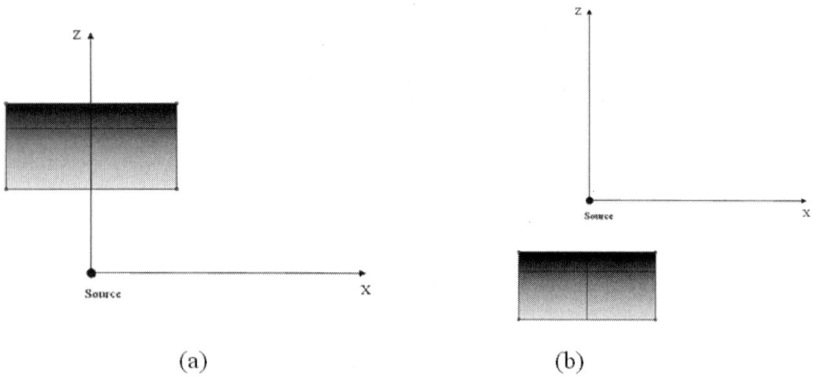

Figure 5.10 Special cases in θ.

5. Once the anxels have been determined, the surface i must be included in the corresponding cells of the AZB matrix that apply the painter's algorithm and consider the other surfaces stored in those cells. The minimum distance from the source to the vertices of the box will be obtained and compared with the minimum distance of the other surface with this aim.

6. The procedure described in steps 1 to 5 will be repeated for all the surfaces of the model.

The AZB matrix created for the direct ray will be used to perform the shadowing test for this ray as well as for the first path (from the source to the corresponding flash point) of all of the effects involved in a UTD analysis.

5.4.2 Application to the Shadowing of an Intermediate Ray

In this case, the objective is to generate an AZB matrix to determine the possible surfaces that can shadow any other path of the ray for any UTD effect, except the first path of the direct ray from the source. Therefore, it will be used to study the shadowing of the intermediate and final paths of all the effects involved in a UTD calculation.

In this case, the AZB matrix is obtained using the following procedure to determine which surfaces are visible from a given one and store them in an AZB matrix:

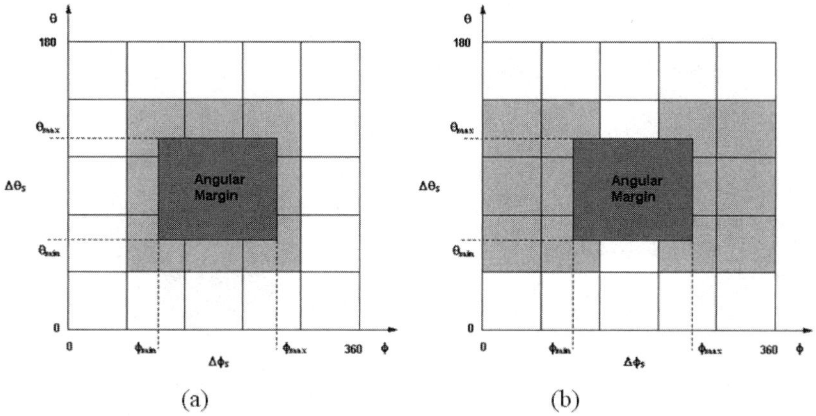

Figure 5.11 AZB matrix with and without special cases.

1. The surface j is selected and its visibility from the surface i is checked applying to the surfaces the criterion of backface culling. A sample of points is performed in each surface and this criterion is applied between a couple of points placed on the surfaces i and j, respectively. If one of the couples satisfies this criterion, the AZB matrix for that surface will be computed. Figure 5.12 shows an example of two surfaces and the application of the criterion to each sample point. The two sample points of each surface $\vec{P}_{ki}$ and $\vec{P}_{lj}$ and the normal vectors $\hat{n}_{ki}$ and $\hat{n}_{lj}$ at each point are used to apply the backface culling criterion twice:

$$\hat{k}_{ij} \cdot \hat{n}_{lj} \geq 0 \tag{5.3}$$

$$\hat{k}_{ij} \cdot \hat{n}_{ki} \geq 0 \tag{5.4}$$

where $\hat{k}_{ij}$ is the direction that starts at the sampling point of the surface i and finishes at the sampling point of the surface j as shown in Figure 5.12.

2. The two bounding boxes for surfaces i and j are determined.

3. Each vertex of the box of the surface i is considered as a source, and the spherical coordinates θ and ϕ of the vertices of the box of j are checked. The special cases described in the previous sections are used to update the spherical coordinates if necessary.

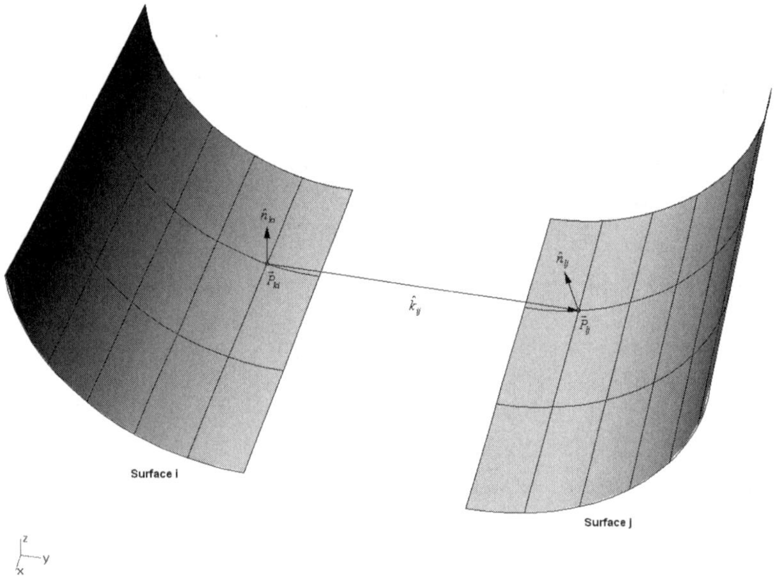

Figure 5.12 Application of the culling criterion to the visibility between surfaces.

4. Once step 3 has been performed, a list of $(\theta, \phi)_{min}$ and $(\theta, \phi)_{max}$ values will be available for each of the eight vertices of the box i that have been considered as sources. From these values the absolute margins of θ and ϕ must be obtained. These absolute margins will determine the rectangle that the surface j will occupy in the AZB plane. The special cases must be reconsidered. Figure 5.13 shows an example in which an absolute margin is obtained from the margins obtained for the two vertices of the surface i. Figure 5.14(a) and Figure 5.14(b) show the angular margin obtained for two vertices and Figure 5.14(c) shows the absolute margin.

5. Once the absolute angular margins have been determined, the anxels of the surface j in the AZB matrix of the surface i are obtained. This surface is stored in the corresponding cells of the matrix.

6. This procedure is repeated for all of the surfaces that are visible from i.

This procedure will be repeated for all of the surfaces of the model. Therefore, each surface will have its own AZB matrix. The advantage of this matrix is that it

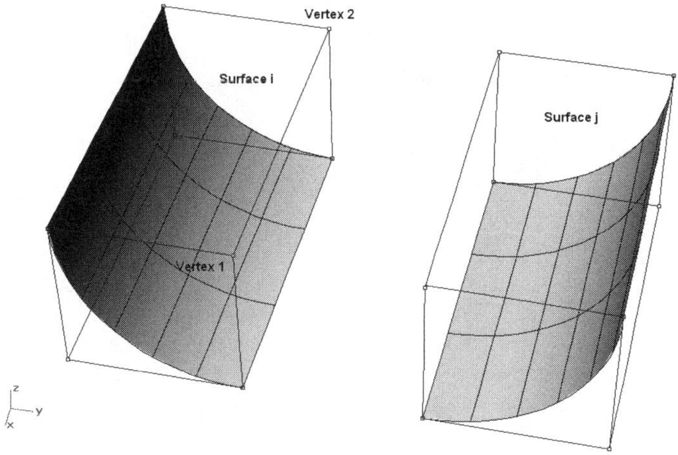

Figure 5.13 Surfaces i and j and their bounding boxes.

only depends on the geometry. For a given geometry, it is possible to create the matrix once, store it in memory, and load it in future analyses of the same geometry.

5.4.3 Application to the Determination of the Reflection Points

The idea here is to determine the potential surfaces that can produce a reflection in a given direction and for a given source. This will help to reduce the number of surfaces in which the minimization process is necessary to obtain a reflection point. The procedure to determine the AZB matrix for the determination of the reflection is as follows:

1. Surface i is a visible surface, and the four vertices of the surface and their normal vectors are used to launch four rays from the surface and apply Snell's law to obtain the direction of reflection of the four rays. A first estimation of the margin of directions or reflection window is obtained. Figure 5.15 shows an example.
2. The four directions are not enough to determine the reflection window in some cases. Therefore, it is necessary to use an adaptive sampling of directions to ensure that all of the reflection directions are considered. This adaptive sampling will be described later. The reflection window will be determined from this sampling. This window will be used to see if a special case appears.

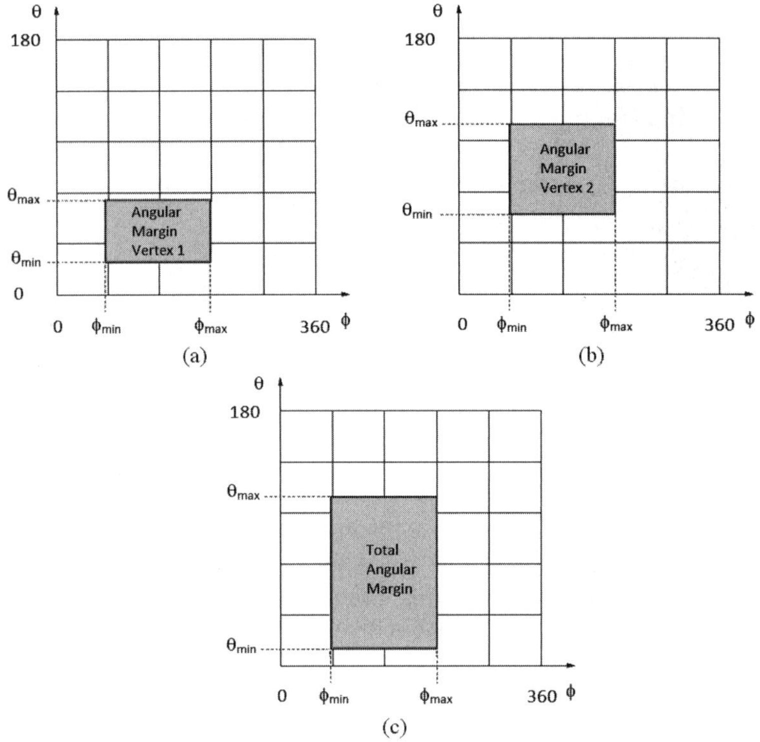

Figure 5.14 AZB matrix with and without special cases.

3. The special cases for θ and ϕ are studied. In the case of θ, it could be necessary to correct the maximum and minimum values to 0° or 180° to determine the values of the cells of the AZB matrix to fill correctly. The four vertices of the surface are used to build two triangular facets that will represent the margin of the directions obtained. Two rays from the origin of coordinates are launched with directions $\hat{z}$ and $-\hat{z}$ to verify that there is an intersection with one of the facets. If there is intersection, the maximum and minimum values of θ must be updated to 180° and 0°, respectively. Figure 5.16 shows an example of this situation. For ϕ, the special cases are identical to the ones described in Section 5.4.1.

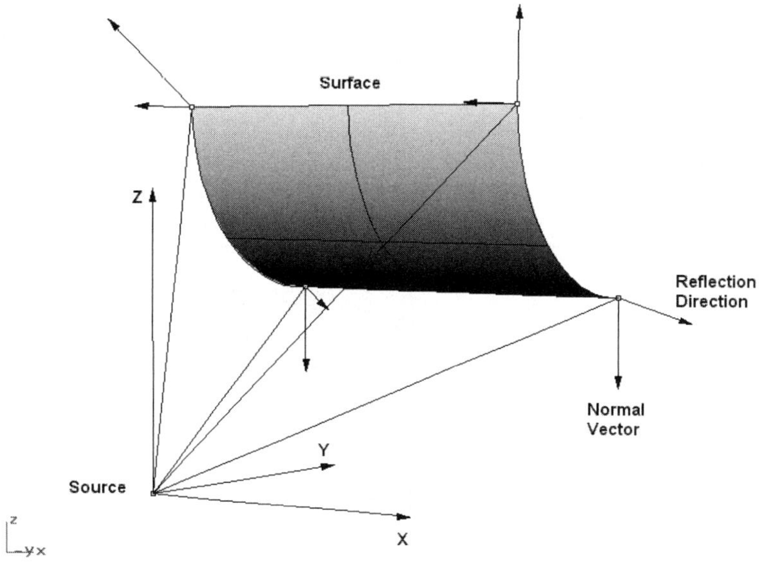

Figure 5.15 First estimation of the reflection window.

4. To make this AZB matrix as efficient as possible, it is convenient to not include the surfaces that are shadowed by other surfaces of the model. To check the shadowing, the four vertices of the surface and the central points are used. If one of the five points is not shadowed, then the surface is considered in the AZB matrix and included in the corresponding cells according to the angular margins obtained in the previous step.
5. This procedure is repeated for all the model surfaces.

5.4.3.1 Adaptive Sampling of the Directions of Surface Reflections

The idea of this sampling is to efficiently obtain all of the possible directions of reflection for a given surface from a given source. As the surface is arbitrarily shaped, it is not enough with the vertices of the surfaces to obtain all of these directions, which happens for plane facets models [10]. It is necessary to have

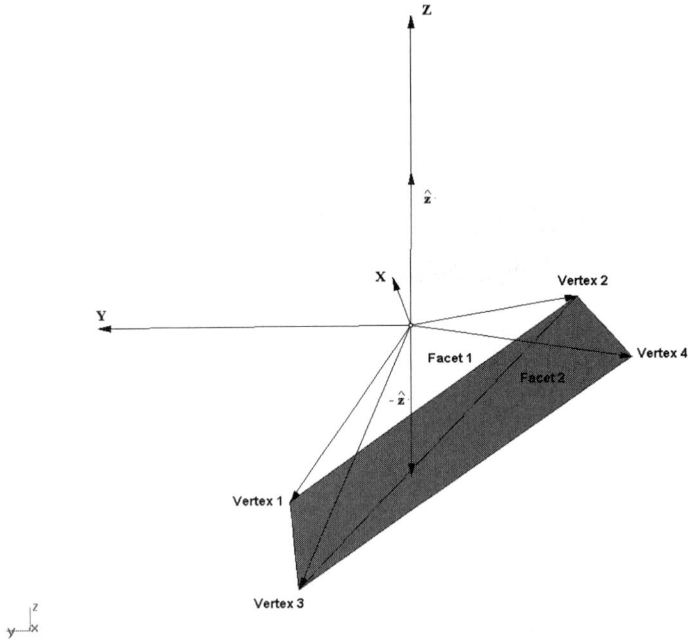

Figure 5.16 An example of the determination of a special case in θ.

an efficient procedure to obtain the complete margin of directions with the lowest computational cost.

This procedure is as follows: the parametric space of the surface is divided into a mesh of $N \times N$ points (see Figure 5.16). Those points define a set of squares formed by four vertices. For each vertex the Cartesian coordinate and the normal vector are obtained. After that, Snell's law is applied to obtain the directions of reflection for each square (θ_i, ϕ_i) with $i = 1, 2, 3, 4$. The reflection window is updated if

$$\theta_{min} < \theta_{i_m in} \text{ or } \theta_{i_m in} > \theta_{max} \tag{5.5}$$

$$\phi_{min} < \phi_{i_m in} \text{ or } \phi_{i_m in} > \phi_{max} \tag{5.6}$$

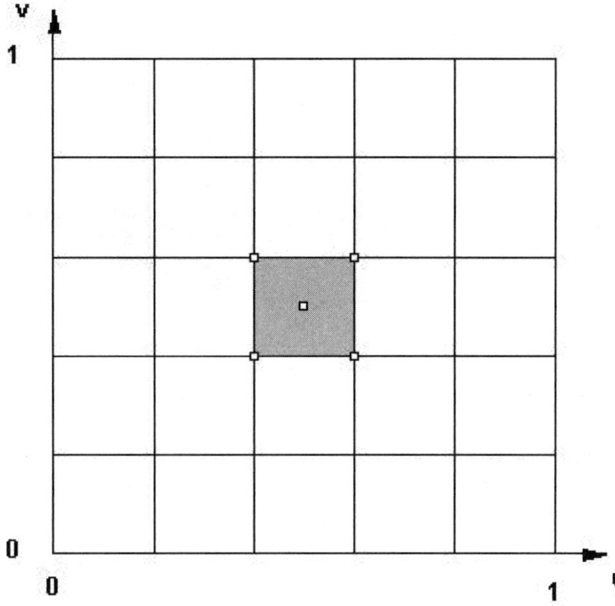

Figure 5.17 An example of sampling of 6x6 points.

Once the direction of reflection of the central point (θ_c, ϕ_c) is associated to each square as shown in Figure 5.17, this direction is checked and contained inside the absolute reflection windows if

$$\theta_{min} \leq \theta_c \leq \theta_{max} \tag{5.7}$$

$$\phi_{min} \leq \phi_c \leq \phi_{max} \tag{5.8}$$

Figure 5.18 shows the correspondence in the real space to the area marked in the parametric space in Figure 5.17.

If this condition is accomplished, the square is marked as complete. If not, this square must be divided into another level of smaller squares until it is complete (or the size of the squares is small enough). This procedure is repeated for all of the squares of the first level.

At the end of the procedure, the complete reflection window will be obtained.

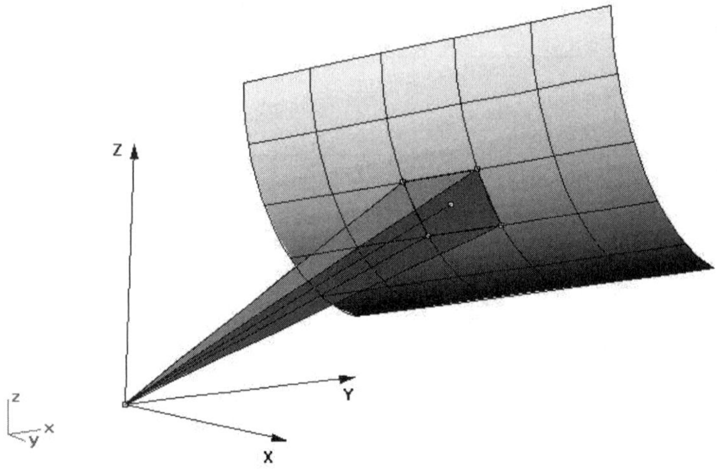

Figure 5.18 The margin in the real space of the area marked in Figure 5.17.

5.4.4 Application to Multiple-Order Reflections

The AZB matrix for multiple-order reflections determines which surfaces produce a reflection after a previous reflection in another surface. The procedure is very similar to the AZB for the simple reflection, but in this case the previous surface is used as the source. Considering i as the illuminated surface where the previous reflection can happen and j as the surface that will be checked, the procedure to create this AZB matrix is as follows:

1. The illumination of j from i must be checked. The eight vertices of the bounding box of surface i will be considered as sources. The illumination will be checked with the four vertices of the surface j using the criterion of backface culling as well as the shadowing test for these points. This shadowing test will use the previous AZB matrices created for the shadowing problem. Figure 5.19 shows an example.

2. If the surface j is illuminated, a reflection window will be created considering the vertices of the box of i as the sources. Eight reflection windows will be obtained. Each reflection window will be obtained using the corresponding adaptive sampling as well as considering the possible special cases.

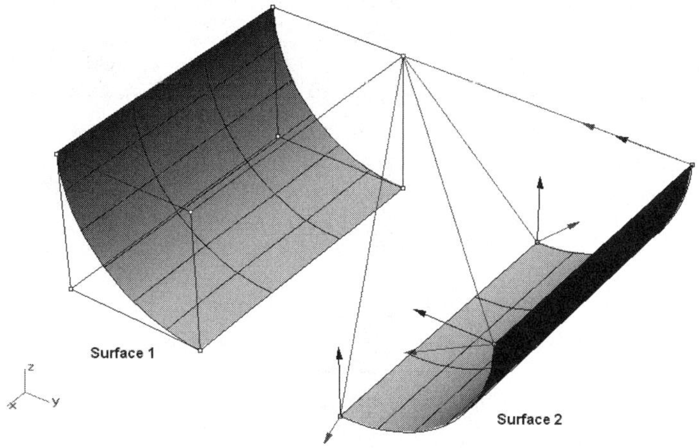

Figure 5.19 An estimation of the angular margin of the vertex of a bounding box.

3. The total reflection window that encloses the eight previous reflection windows is obtained. The cells of the AZB matrix to fill are determined from it.
4. This procedure is repeated for all of the surfaces illuminated from the surface i.
5. This procedure is repeated for all of the model surfaces. Every surface has its own AZB matrix reducing the computational time associated with a multiple reflection where that surface is involved.

This algorithm only depends on the geometry. Therefore, it is only necessary to perform it once for a given geometrical model. This information can be used to generate a reflection window for a multiple reflection by following the same procedure for a simple reflection.

5.4.5 Application to the Shadowing of the Diffracted Ray

The application to the diffracted ray for curved surfaces is analogous to the case of straight edges [10]. In this case, the coordinates β and α are used instead of the spherical coordinates θ and ϕ. These coordinates are defined as follows: β is the angle of the Keller's cone for a diffraction point, and α is the angle formed by the diffracted ray and the first surface of the wedge (see Figure 5.20). These coordinates accomplish a surface arrangement. For each source-edge pair, the space where diffracted rays exist is limited by the edge coordinates $(\beta_{min}, \beta_{max}, \alpha_{min}, \alpha_{max})$. The edge coordinates fix the margins of the matrix that is known as the AZB matrix of diffraction. This matrix is divided into anxels. The AZB matrices for diffraction are created in the same way as those used for reflection, but the new coordinates β and α are considered during the creation of AZB matrices.

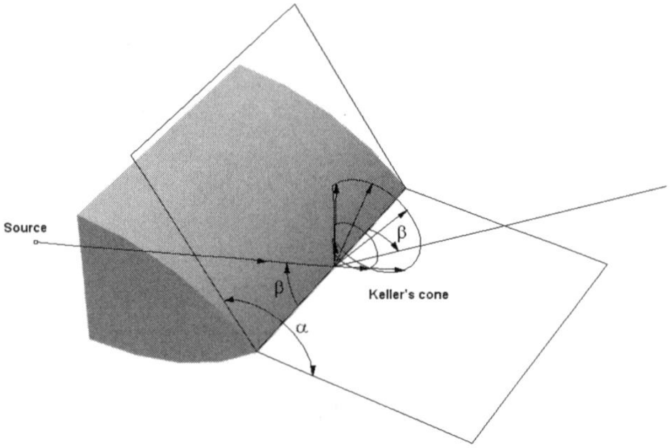

Figure 5.20 Angles β and α for the AZB of diffraction.

5.5 COMBINATION OF AZB-SVP FOR THE ANALYSIS IN THE NEAR ZONE

If the analysis is performed in the near zone, it is necessary to consider the direction of observation as well as the position of the observation points. This type of analysis is especially suitable for certain problems as the propagation in mobile communications. An example of this kind of problem will be shown in Chapter 6.

The SVP is based on the division of the space in voxels. In this case, the space covered contains the observation points in which the field must be obtained. Figure 5.21 shows an example.

Once the division has been accomplished, the AZB matrices for the calculation of the simple reflection and for the multiple reflections can be created. The procedure is the same for the directions of observation, but an additional condition

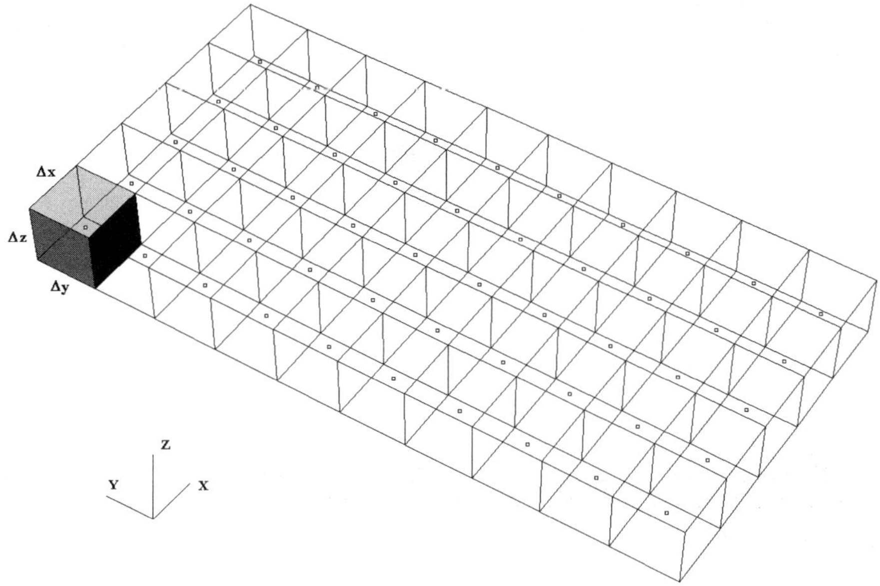

Figure 5.21 Division of the observation points in voxels.

must be accomplished. This condition is to determine which voxels of the model are included in the corresponding reflection window. Therefore, only the observation points contained on those voxels will be considered during the analysis of the corresponding effect.

If a voxel can be reached from a reflection coming from surface i, that voxel must be included in the reflection window of the surface as shown in Figure 5.22. This inclusion may be determined by obtaining the spherical coordinates of the eight vertices of the voxel with respect to the vertices of the bounding box of the surface i. The corresponding special cases must also be considered. The minimization for obtaining the reflection points will only be performed for the observation points contained in the voxels that are included in the reflection window of the surface.

The same idea is used to include a voxel in the reflection window for a multiple reflection case. It is important to notice that the empty voxels (voxels without observation points) will not be considered in the analysis.

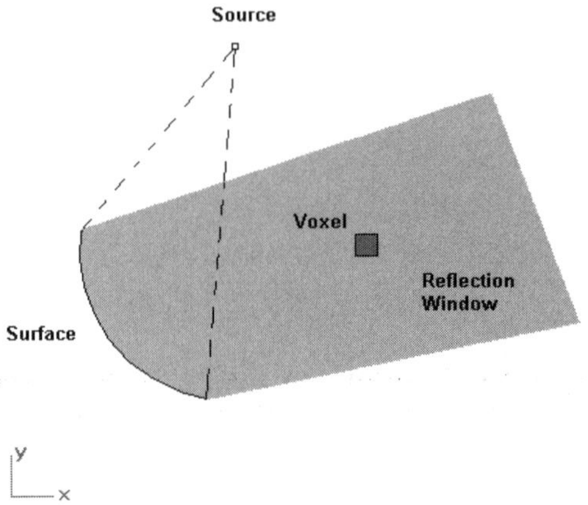

Figure 5.22 The inclusion of a voxel in the reflection window.

5.6 APPLICATION OF AZB FOR THE ANALYSIS OF THE RCS

5.6.1 Application to Plane Facets

In the case of the RCS, when the incident wave is plane, the contribution to the monostatic RCS of a plane facet is maximized when the direction of scattering is parallel to the normal vector to the surface and its value becomes negligible when this direction is far from the normal. In a complex model, a single facet is considered as a potential contributor only when the direction of scattering is close to the normal vector. The margin must be selected carefully following a compromise between the accuracy of the results and the computational time.

The selection of an angular region in which the facet contributes allows creating an AZB matrix that corresponds with the margin of directions in which the facet can contribute to the RCS. The selected margin will define a reflection window with some values of $(\theta_{min}, \theta_{max})$ and (ϕ_{min}, ϕ_{max}). The facet will be stored in the anxels contained in this reflection window. Repeating this method for each facet of the model creates the AZB for the reflection by plane facets. Figures 5.23 and 5.24 show examples of the angular margin for a given plane facet together with the cells that this facet occupies in the AZB matrix.

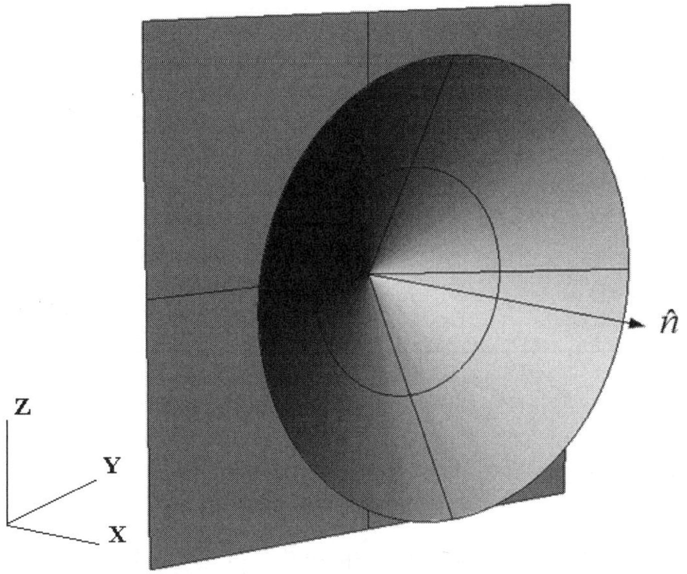

Figure 5.23 The angular margin is where the monostatic RCS contribution of a plane facet is significant.

For a plane wave incidence, the same special cases can appear for a spherical wave incidence. These factors can be considered when the AZB matrix is completed.

5.6.2 Application to Curved Surfaces

The procedure for curved surfaces is analogous to the procedure for plane facets. In this case, there is not a single normal vector for the entire surface because the normal vector is a function of the parametric coordinates. Therefore, the procedure of the plane facets is repeated for the normal vectors of a sample of points over the surface. Each point will define a reflection window. The union of these reflection windows will constitute the total reflection window of the surface in a procedure analogous to the procedure used for a spherical wave. To ensure that all of the directions of the

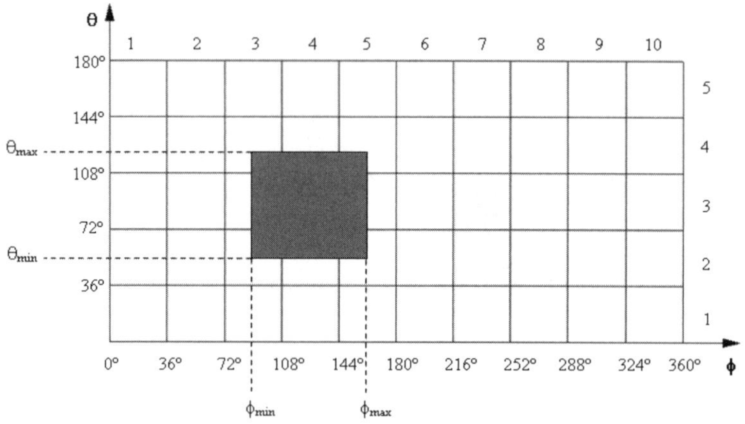

Figure 5.24 An inclusion of the facet in the AZB matrix.

reflection window are included, an adaptive sampling as proposed in Section 5.4.3 is advisable.

5.6.3 Application to Double- and Higher-Order Effects Between Plane Facets

As in the simple reflection, there is an angular margin where the multiple-order reflection between the plane facets contributes in a significant way to the RCS. The condition to determine this margin is quasi-perpendicularity. Figure 5.25 shows an example of this condition for a double reflection.

In Figure 5.25, the condition of quasi-perpendicularity is expressed mathematically with the following expressions:

$$\cos\left(\frac{\pi}{2} - \alpha_2\right) \leq \hat{k}_s \cdot \hat{n}_1 \leq \cos\left(\alpha_2\right) \tag{5.9}$$

$$\cos\left(\frac{\pi}{2} - \alpha_2\right) \leq \hat{k}_s \cdot \hat{n}_2 \leq \cos\left(\alpha_2\right) \tag{5.10}$$

$$\left|\hat{e} \cdot \hat{k}_s \hat{n}_2\right| \leq \cos\left(\frac{\pi}{2} - \alpha_1\right) \tag{5.11}$$

where α_1 is the angular margin in which the scattering direction is quasi perpendicular, α_2 is the angular region to eliminate the directions with grazing incidence to

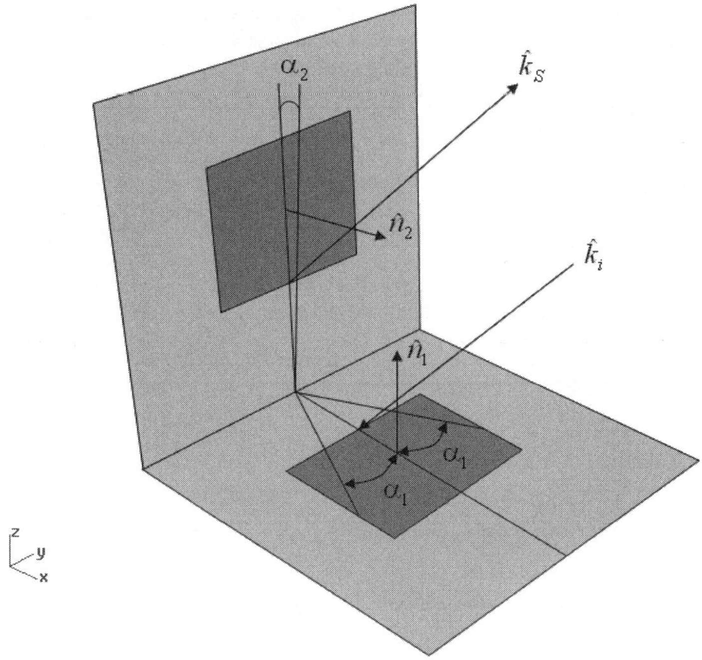

Figure 5.25 The condition of quasi-perpendicularity.

the surface, and $\hat{e}$ is the unit vector of the segment that intersects two planes that contain both surfaces (the edge between the two surfaces).

Once the margin angular has been fixed, this margin determines the spherical coordinates in which a given couple of facets produce double reflection. These coordinates can be stored in an AZB matrix as in the cases of the previous sections.

References

[1] A. S. Glassner. *An Introduction to Ray Tracing*. Academic Press, 1989.

[2] J. D. Foley, A. Van Dam, S. K. Feiner, and J. F. Hughes. *Computer Graphics. Principles and Practice*. Addison-Wesley, 1995.

[3] M. F. Cátedra, J. Pérez, F. Saez de Adana, and O. Gutiérrez. "Efficient ray-tracing techniques for three-dimensional analyses of propagation in mobile communications: Application to picocell and microcell scenarios." *IEEE Antennas and Propagation Magazine*, 40(2):15–28, April 1998.

[4] F. Saez de Adana, O. Gutiérrez, I. González, J. Pérez, and M. F. Cátedra. "Propagation model based on ray tracing for the design of personal communication systems in indoor environments." *IEEE Transactions on Vehicular Technology*, 49(6):2105–2112, November 2000.

[5] G. Liang and H. L. Bertoni. "A new approach to 3D ray tracing for propagation prediction in cities." *IEEE Transactions on Antennas and Propagation*, 46(6):853–860, June 1998.

[6] F. Aguado Agelet, A. Formella, J. M. Hernado Rábanos, F. Isasi de Vicente, and F. Pérez Fontán. "Efficient ray-tracing acceleration techniques for radio propagation modeling." *IEEE Transactions on Vehicular Technology*, 49(6):2089–2104, November 2000.

[7] Z. Yun, M. F. Iskander, and Z. Zhang. "Fast ray tracing procedure using space division with uniform rectangular grid." *Electronics Letters*, 36(10):895–897, May 2000.

[8] Z. Yun, Z. Zhang, and M. F. Iskander. "A ray-tracing method based on the triangular approach and application to propagation prediction in urban environments." *IEEE Transactions on Antennas and Propagation*, 50(5):750–758, May 2002.

[9] H. Fuchs. "On visibility surface generation by a priori tree structures." *Computer Graphics*, 14(3):124–133, July 1980.

[10] M. F. Cátedra and J. Pérez Arriaga. *Cell Planning for Wireless Communications*. Artech House, 1999.

[11] H. S. Rubin and T. Whitted. "Three-dimensional representation for fast rendering of complex scenes." *Computer Graphics, Proc.*, pp. 110–116, 1980.

[12] F. Aguado Agelet and A. Formella. "Fast ray tracing for microcellular and indoor environments." *IEEE Transactions on Magnetics*, 33(2):1484–1487, March 1997.

[13] M. Kaplan. "Space-tracing: A constant time ray-tracer." *Course Notes from Tutorial State of the Art in Image Synthesis. SIGGRAPH 85*, 1985.

[14] D. N. Schetino and F. J. S. Moreira. "A ray-tracing algorithm based on a double-tree image theory." *IEEE MTT-S IMOC Proceedings*, pp. 199–202, 2001.

[15] Z. Zhang, Z. Yun, and M. F. Iskander. "3D tetrahedron ray tracing algorithm." *Electronics Letters*, 37(6):6–15, March 2001.

[16] C. A. Balanis. *Advanced Engineering Electromagnetics*. John Wiley and Sons, 1989.

[17] J. Pérez, F. Saez de Adana, O. Gutiérrez, I. González, M. F. Cátedra, I. Montiel, and J. Guzmán. "FASANT: Fast computer tool for the analysis of on board antennas." *IEEE Antennas and Propagation Magazine*, 41(2):94–98, April 1999.

[18] E. A. Hines and D. P. Greenberg. "The light buffer: A shadow-testing accelerator." *IEEE Computer Graphics and Aplications*, 6(9):6–16, September 1986.

[19] M. E. Newell, R. G. Newell, and T. L. Sancha. "A solution to the hidden surface problem." *Proceedings of the ACM National Conference*, pp. 443–450, 1972.

Chapter 6

Applications

6.1 OVERVIEW OF ASYMPTOTIC TECHNIQUE CODES

In this chapter, several examples of the application of the techniques to real-world problems described in the previous chapters are described. Specifically, the results of the application of GO/UTD and PO/PTD are presented. The results were obtained with two programs: FASANT [1, 2], which is based on GO/UTD and is applicable to problems of radiation and propagation such as on-board antennas or mobile communications channel characterization, and POGCROS [3, 4] which is based on PO/PTD, for which the main application is RCS computation. Demo versions of both codes and their user manuals are included in the DVD accompanying this book. These codes are used because they were developed by the authors of this book, and therefore, the practical considerations of analyzing such complex cases with those codes can be readily included in the book.

However, FASANT and POGCROS are not the only commercially available codes based on asymptotic techniques. Therefore, it is interesting to mention some of the tools that can be used by an engineer to solve electromagnetic problems. With this purpose some codes are cited here. FEKO [5], developed by EM Software and Systems, combines rigorous techniques, such as the method of moments (MoM) and the finite element method (FEM), with asymptotic techniques like UTD and PO for the analysis and design of antennas, the calculation of the RCS, the placement of on-board antennas, and the analysis of problems of electromagnetic compatibility (EMC). SARGASSES [6] was developed by Mothesim in France and is based on UTD with the purpose of the analysis of on-board antennas. XGTD [7], developed by REMCOM, is also based on UTD and is devoted to the analysis of the interaction of antennas mounted on complex structures and implementing ray-tracing based on the SBR method [8]. SUPERNEC [9], which belongs to Poynting Software, implements a hybrid method that combines MoM with UTD for antenna design, analysis, EM consulting work, and antenna placement studies. Antenna design framework (ADF) [10], developed by IDS under an European Space Agency (ESA) contract, includes a set of modeling tools based on MoM, PO, and UTD for the analysis and design of antennas including the surrounding structure influence. GRASP [11], developed by TICRA for the analysis of reflector antennas and arrays, includes modules such as the one based on UTD for the analysis of the influence of complex structures over the antenna radiation pattern. XPATCH [12] for RCS computation was discussed in Chapter 4. RCSSIG [13] was developed by SAIC, and is based on PO/PTD with the aim of facilitating the RCS analysis of complex 3-D targets including aircrafts, land vehicles, and ships. This list of codes is not complete, and it is an attempt to provide the reader with an overview of some of the most interesting tools that can be used for the analysis of real engineering problems using asymptotic techniques.

This chapter is divided into two sections: the results obtained with the GO/UTD technique and the results obtained with PO/PTD. In the first case, FASANT code was used, and in the latter case, POGCROS was used, except in the last example. In this case, the results were obtained by solving the PO integral numerically. In each section, several cases analyzed with the aforementioned programs are included. Each case includes a description of the problem and the results obtained with the programs, along with a discussion of its application to practical problems. Some of the results shown are compared with measurements, when available, or other methods belonging to the category of rigorous methods, such as MoM [14]. In particular, MONURBS, a computer code based on MoM with a multilevel fast multipole [15, 16] formulation, was used to prove the validity for the asymptotic approaches presented throughout this book. This code is also included in the companion DVD. As with FASANT and POGCROS, the reason for using MONURBS was that it has been developed by this book's authors and can therefore be readily incorporated into the text.

6.2 RESULTS OBTAINED WITH FASANT

FASANT is a computer tool that combines the GO/UTD formulation presented in Chapter 3 with the ray-tracing acceleration techniques presented in Chapter 5. Internally, the geometry analyzed by FASANT is modeled by means of NURBS and planar surfaces. FASANT inputs geometrical files in DXF, SAT, and IGES formats. Thus, the geometry can be created with most commercially available CAGD programs. Also, the program has a graphical interface for creating new geometries and modifying existing ones. Other capabilities of the interface are the visualization of the ray-tracing, obtaining the ray information as ray trajectories, comparison of the results, and others; these can be seen on the accompanying DVD.

The functionalities of FASANT are the computation of the radiation pattern in the far field, the computation of the field at observation points near the structure, the determination of coupling between antennas, and the computation of representative parameters of the propagation channel. It is possible to choose the number of interactions to take into account when these tasks are performed, considering as interactions a single reflection or diffraction. Therefore, if two interactions are chosen, the program calculates the combination of reflections and diffractions of the second-order that influence the final result.

This section presents four GO/UTD application examples using the FASANT tool. The first two examples use GO/UTD to calculate the radiation patterns of antennas on board complex structures. The third example deals with propagation in urban areas combining indoor and outdoor propagation. Specifically, in this example, indoor coverage from satellite propagation is predicted using FASANT. Finally, the fourth example shows the application of FASANT for determining the interference in radio systems created by wind turbines located in their vicinity.

6.2.1 Radiation Pattern of Antennas On-Board the Jason and Stentor Satellites

One fundamental feature to be characterized when designing communication satellites is the influence of the structure on on-board antennas. The main parameters of the antenna, such as the radiation pattern or the cross-polar polarization level, are affected by the structure. Using simulation tools is a good way to reduce costs in the process of optimizing the location of the antennas. As the dimensions of the satellites are large compared to the wavelengths of most communication frequencies, asymptotic techniques and, more specifically, GO/UTD are very suitable for this kind of analysis.

The following sections present example applications of the analysis of on-board satellite antennas using the FASANT code. In these examples, the antennas on board the Jason and Stentor satellites were considered and the results compared with previously published measurements; the FASANT results have been reported previously [17–19].

6.2.1.1 Jason Satellite

This satellite was designed by NASA and the French Centre National d'Études Spatiales (CNES) with the objective of producing global maps of ocean-surface topography. Before the installation of the antennas on board the satellite, a mock-up of a part of it was used to test the influence of the structure on their radiation pattern (see Figure 6.1). At the bottom, the position of the antenna to be measured can be seen. The frequency of operation of the antennas was 2.2 GHz, and the structure dimensions were 2.22m $\times$ 1.158m $\times$ 1.507m. In this example the radiation pattern of the antenna on board the satellite will be computed using FASANT code.

The first step is to create the geometrical model. This task can be performed starting from the satellite planes using any number of available CAGD programs such as AUTOCAD, Microstation, Catia, and Rhinoceros, among others. Once the geometry has been created, it must be exported to a file containing the necessary geometrical data for the application of the GO/UTD method. FASANT can input SAT and IGES formats. These formats are available in most of the CAGD programs. Internally, FASANT code uses the DXF format, and FASANT handles two kinds of primitives, one for plane surfaces (3DFACE) and the other for curves (POLYLINE).

The geometry introduced to FASANT is created based on the satellite dimensions. An important feature to take into account when making the geometric model is that details that have a smaller size than the wavelength should not be modeled because of GO/UTD restrictions. Figure 6.2 shows the geometrical model created for this case. Comparing the real satellite mock-up (Figure 6.1) with the model (Figure 6.2) shows slight differences. For instance, the top mock-up is modeled with a single plane facet without taking into account flanges and screws. Another dissimilarity is the helical antenna on the right side of the parabolic reflector. This antenna has been modeled using a cylinder and a cone in which the cylinder dimensions match the spiral width and height.

Another input parameter to take into account is the antenna itself. The physical antenna model is not created (Figure 6.2), but rather the antenna is substituted by its radiation pattern in free space. This can be measured in an anechoic chamber or can be computed using analytical or rigorous methods, such as MoM and FDTD, among others. In this case, the measurements of this radiation pattern were provided by CNES. In FASANT, there are three different file formats that describe the antenna:

- REV, when the radiation pattern has a symmetry of revolution (the Z axis is assumed to be the revolution axis);

- RV2, when the radiation pattern is defined by E-plane and H-plane patterns (again the Z axis is forced to be the axis of symmetry);
- 3DE, when the radiation pattern is three-dimensional.

An example of each of these files is included on the book's companion DVD. In the case of the antenna described here, the second kind of radiation pattern (RV2) was used because the antenna was described by its E and H planes. Figure 6.3 shows the measured antenna radiation pattern for these planes in which the E plane is in the XZ plane of the antenna, and the H plane is the YZ plane. The antenna position of the phase center and the antenna orientation are shown in Figure 6.2.

Figure 6.1 Jason satellite mock-up.

Once geometry and antenna parameters have been determined, the rest of the input parameters must be fixed. These parameters are:

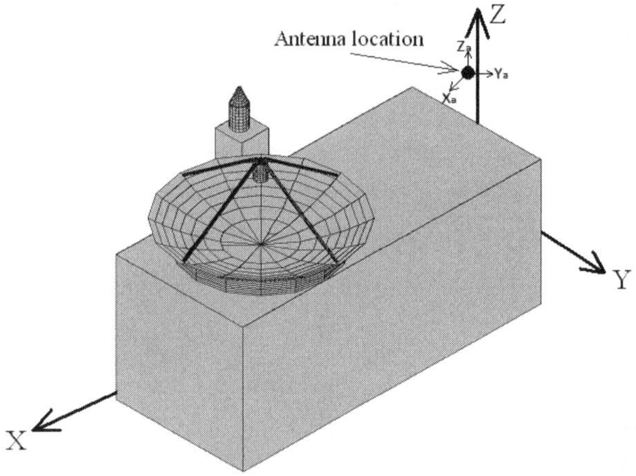

Figure 6.2 Geometric model and antenna position.

- Type of analysis: near or far field. In the first case, the observation points in which the radiated field will be computed must be specified. In the second case, the observation directions of the desired radiation pattern must be introduced. The objective was to determine the structural influence on the radiation pattern. With this aim, the radiation pattern of the on-board antenna was computed in several directions, specifically, in three cuts ($\theta = 45°, 90°$, and $135°$), with ϕ varying from $-90°$ to $90°$.
- Sweep in frequency: in this example, the radiated field is computed only at one frequency (2.2 GHz).
- Interactions: this parameter defines the order of interactions to take into account in the simulation for computing of the field. An interaction is a reflection or diffraction in the structure. The order must be selected depending on the geometrical model and the angular sweep to be analyzed. If a high order of interaction is selected, the accuracy of the result will improve, but the computational time will increase exponentially. If the order is low, the computational time will decrease, but the result may not be accurate. In this case, two interactions have been considered. Given the structure of the satellite, a third interaction would not significantly improve the accuracy of the results.

Figures 6.4 to 6.9 show the GO/UTD results for the specified directions compared to the measured values. The measurements were performed in an anechoic chamber in the CNES laboratories. Both results and measurements were obtained in circular components [20]. As we saw in Chapter 3, UTD expressions obtain the radiated field in linear components ($\vec{E}_\theta$ and $\vec{E}_\phi$). Circular components can be calculated easily from the linear components:

$$\vec{E}_r = \frac{\vec{E}_\theta - j\vec{E}_\phi}{\sqrt{2}} \tag{6.1}$$

$$\vec{E}_l = \frac{\vec{E}_\theta + j\vec{E}_\phi}{\sqrt{2}} \tag{6.2}$$

where $\vec{E}_r$ is the right component and $\vec{E}_l$ is the left component. Figures 6.4 to 6.9 show the copolar (left) and cross-polar (right) components.

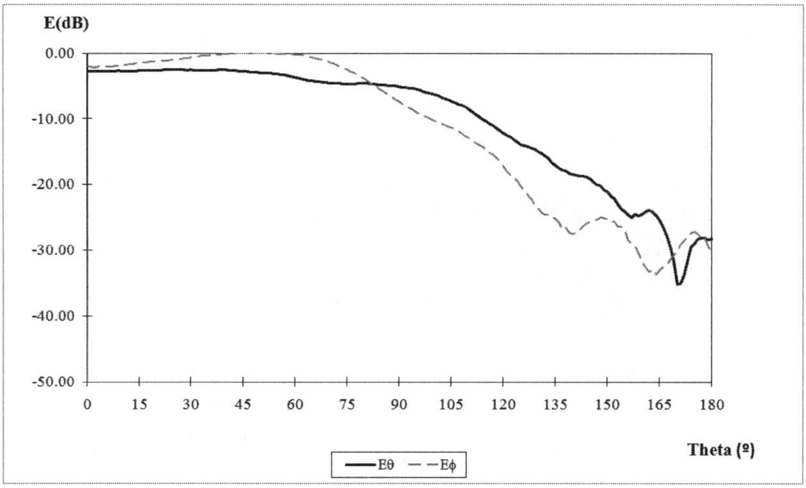

Figure 6.3 Radiation pattern of the antenna in free space.

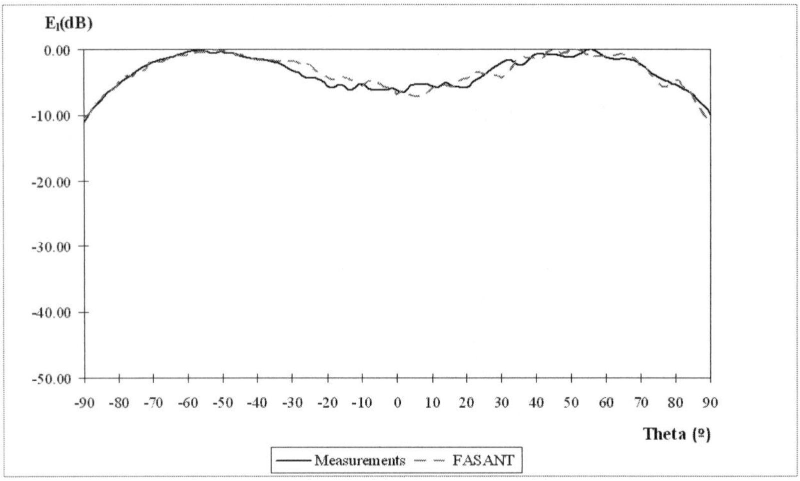

Figure 6.4 E_l components for $\phi = 45°$.

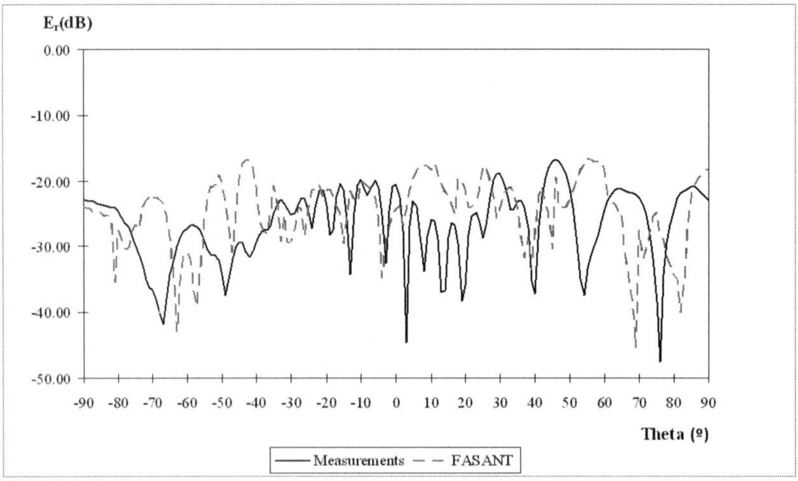

Figure 6.5 E_r components for $\phi = 45°$.

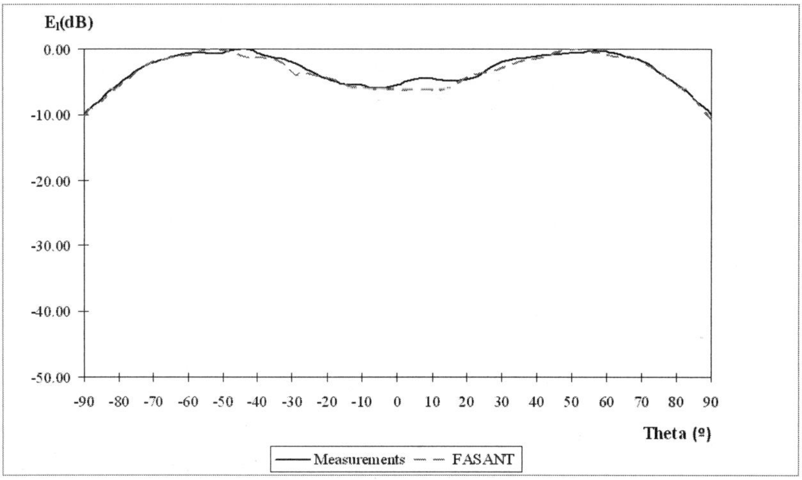

Figure 6.6 E_l components for $\phi = 90°$.

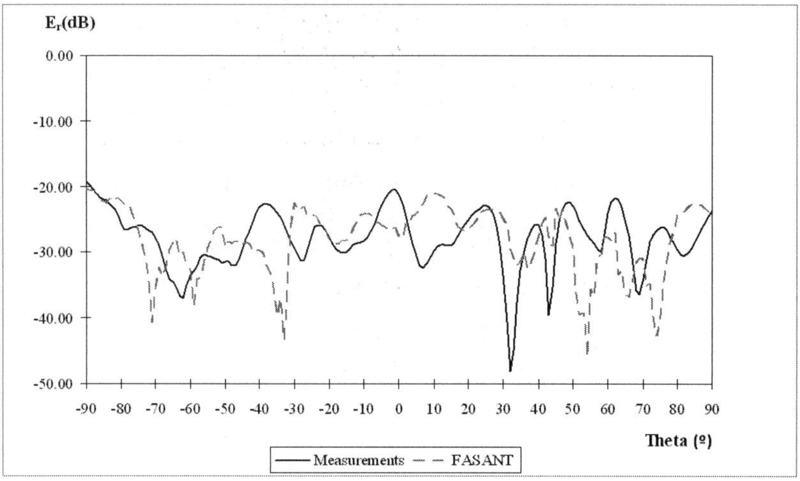

Figure 6.7 E_r components for $\phi = 90°$.

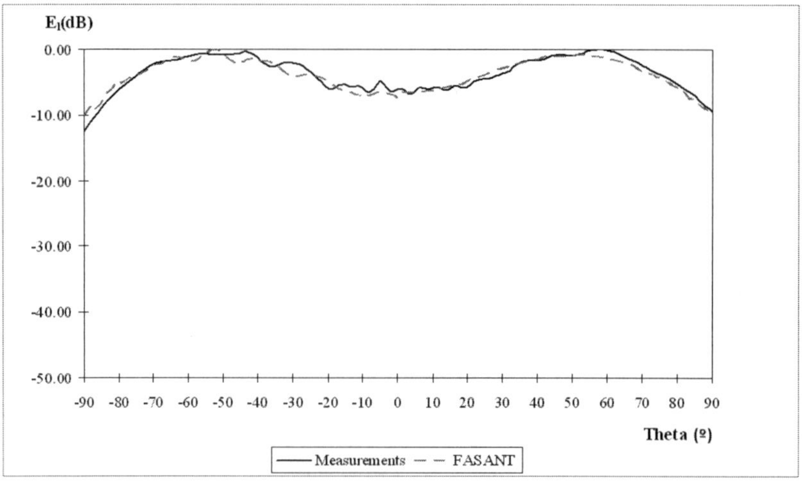

Figure 6.8 E_l components for $\phi = 135°$.

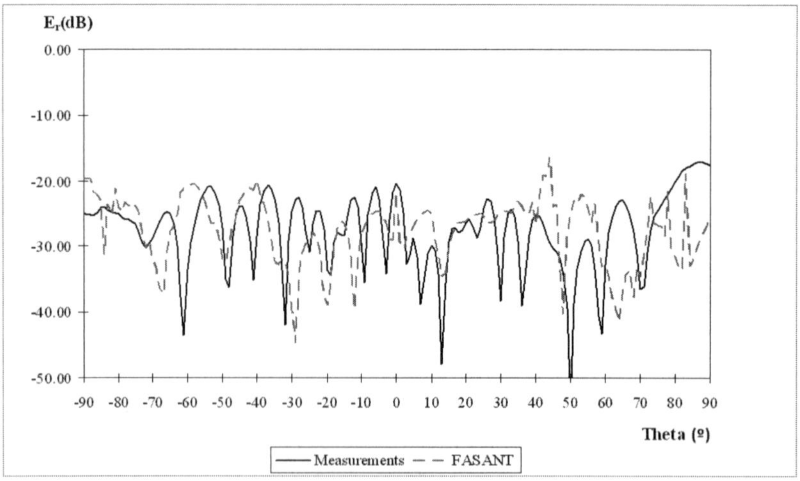

Figure 6.9 E_r components for $\phi = 135°$.

A good agreement between predictions and measurements is observed for the copolar polarization. While the cross-polar polarization values differed, the predicted and measured levels for this component were quite similar. These differences may be due to several elements of the satellite that do not satisfy the requirements of GO/UTD because they are small in comparison with the wavelength. As the cross-polar polarization is between 20 and 30 dB lower than the copolar polarization, the radiated field is contaminated with the noise level, and these small elements produced differences in the predicted levels.

6.2.1.2 Stentor Satellite

Stentor (Satellite de Telecommunications pour Experimenter les Nouvelles Technologies en Orbite) is part of a French satellite program with the objective of proving the capabilities of emergent technologies and of performing flight qualification of some new equipment. The same study performed for the Jason satellite was performed for the Stentor using a mock-up built in the CNES laboratories, shown in Figure 6.10.

Figure 6.10 Stentor satellite mock-up.

The geometrical model of the Stentor is built following the same process discussed in Section 6.2.1.1. In the case of the Stentor, the structure is more complex. It is composed of 83 planar surfaces and 11 curved surfaces (Figure 6.11). The dimensions of the structure are $1.3 \times 0.8 \times 0.5$ meters. The transmitter is a pyramidal GPS (Figure 6.12) located at the position indicated in Figure 6.11. Figure 6.13 shows the radiation pattern of the antenna in free space at its operating frequency of 1.575 GHz.

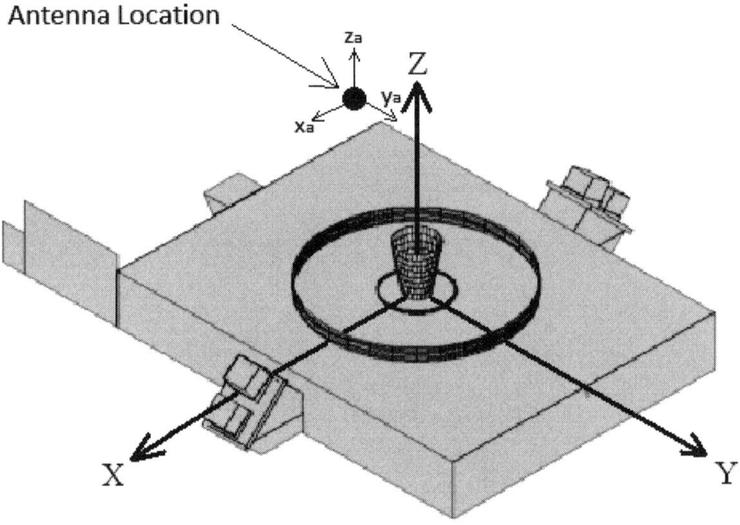

Figure 6.11 Geometrical model of the Stentor satellite.

Figures 6.14 to 6.19 show the circular components, E_r and E_l, of the radiated fields of the on-board antennas for the cuts at $\theta = 45°$, $90°$, and $135°$.

The analysis of the results in this case showed patterns very similar to those for the Jason satellite.

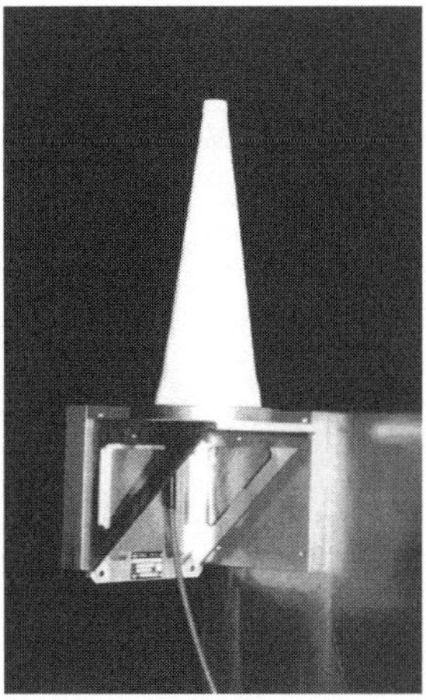

Figure 6.12 Pyramidal GPS antenna.

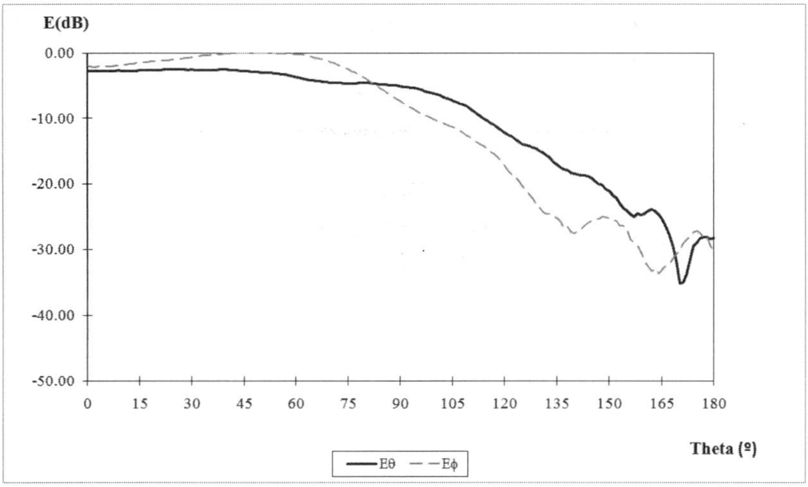

Figure 6.13 GPS antenna radiation pattern in free space.

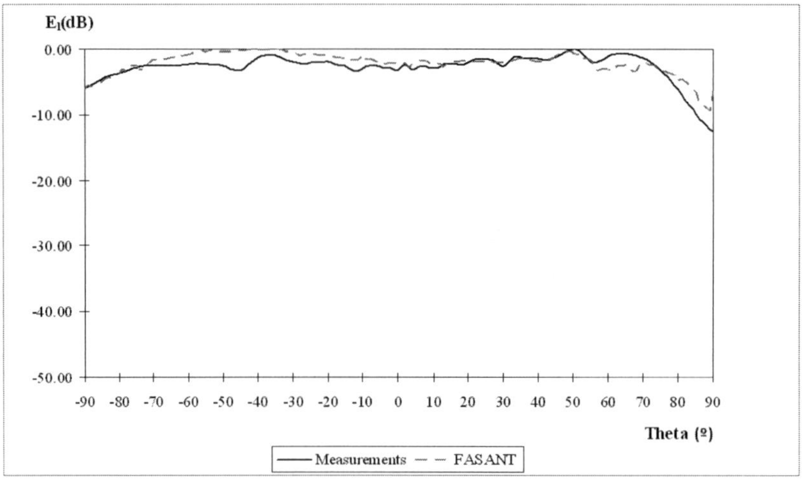

Figure 6.14 E_l components for $\phi = 45°$.

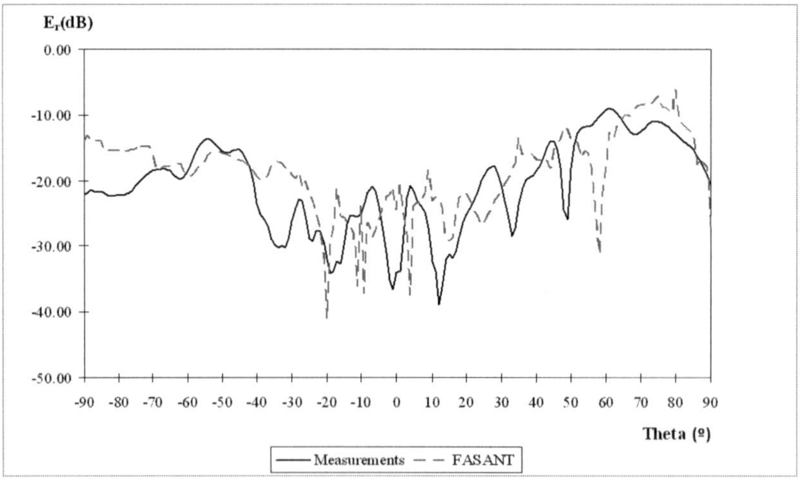

Figure 6.15 E_r components for $\phi = 45°$.

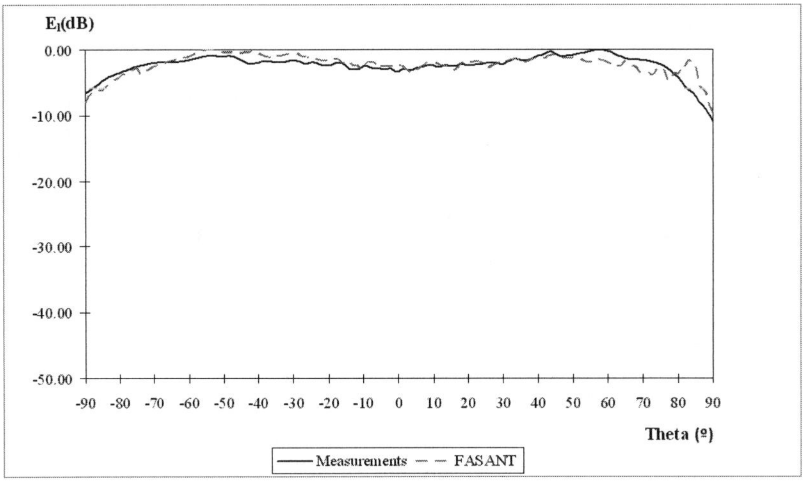

Figure 6.16 E_l components for $\phi = 90°$.

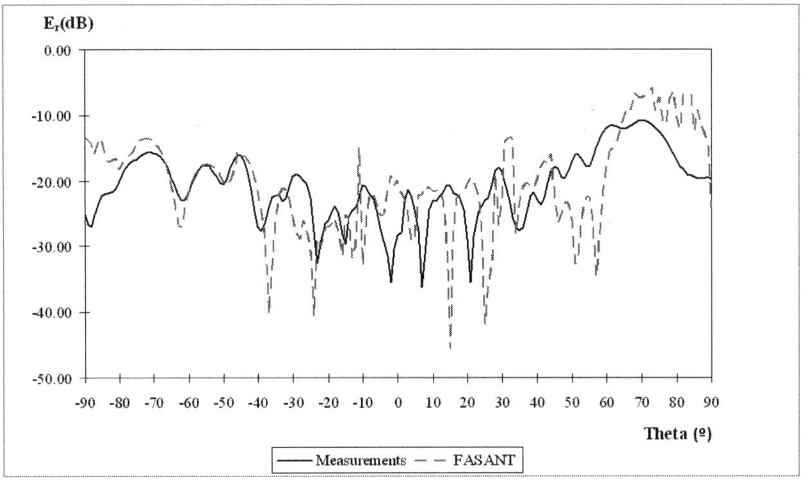

Figure 6.17 E_r components for $\phi = 90°$.

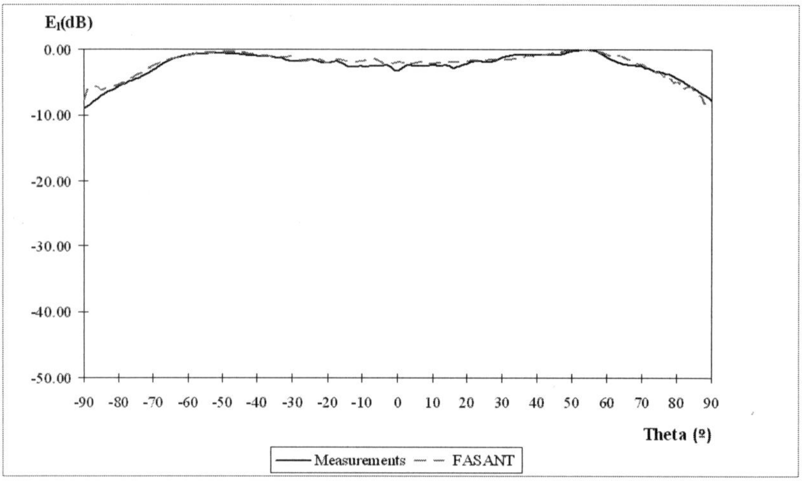

Figure 6.18 E_l components for $\phi = 135°$.

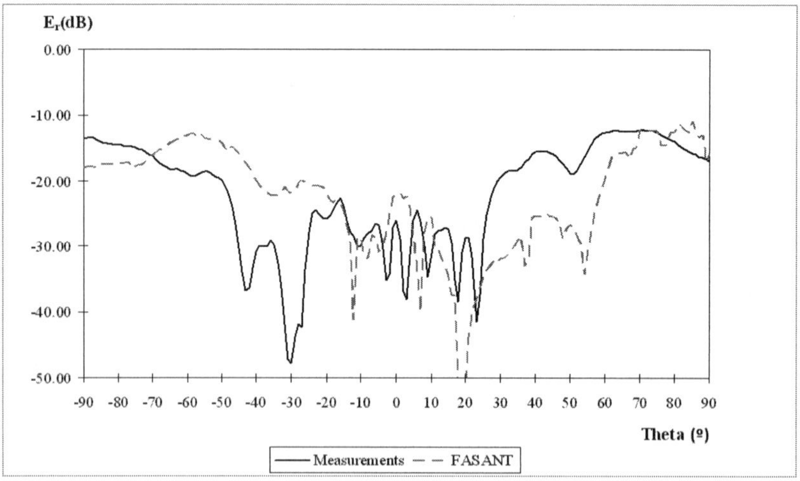

Figure 6.19 E_r components for $\phi = 135°$.

6.2.2 Radiation Pattern of Antennas On-Board the International Space Station (ISS)

The ISS is the largest artificial satellite orbiting the Earth. It is used as a research laboratory for applications in the fields of biology, physics, astronomy, and others. One of the projects on the ISS is the Sport project, designed and funded by the Italian Space Agency to be the first scientific microwave polarimeter for space astrophysics. The objectives of this project are the construction of precise maps of the galactic polarized emissions at 22 and 32 GHz and the measurement of the polarizations of the cosmic microwave background radiation at 60 and 90 GHz. The antennas for this project have been designed with particular characteristics to accomplish these requirements. As mentioned in the previous section, as in all the satellite installations, the placement of these antennas required a preliminary study of the influence of the structure of the ISS on the antenna features. A project funded by the European Space Agency (ESA) was initiated to perform a preliminary study considering the following aspects:

- The possible degradation of the radiation pattern of a representative antenna for several positions of the solar panels. These panels are mobile structures that are continually pointing toward the sun. Therefore, the panels have different positions with respect to the structure throughout the day. For this reason, there may be positions of the panels that interfere with the radiation pattern. It was thus necessary to study several positions of the panels and calculate the radiation patterns for each of those positions to find the critical cases.
- An analysis of the degradation of the cross-polar radiation level of the antenna for two critical positions of the solar panels. According to the antenna specifications, the cross-polar component must be below -45 dB. The structural influence, especially the diffraction by the panels, could increase the cross-polar component, and this increment was studied.

Following the same process as for the previous cases, a CAGD model of the station was created. This example requires different geometries to be analyzed with different solar panels positions. Once one ISS geometry is created, it is possible to obtain the rest of the configurations by changing only the surfaces corresponding to the solar panels. Three different configurations were used to compute the radiation pattern of the antenna; the results have been published previously [21, 22].

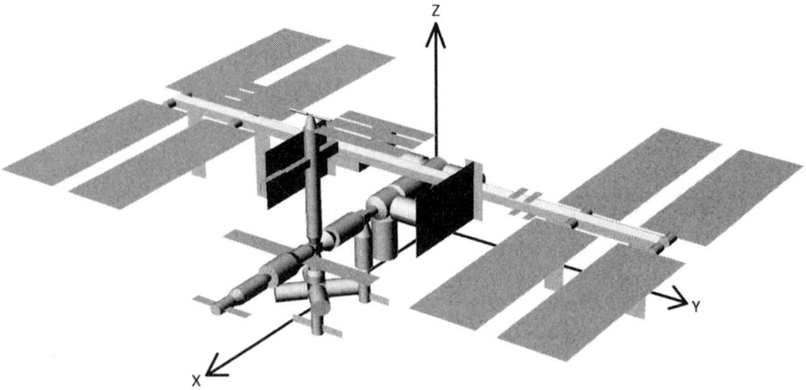

Figure 6.20 ISS configuration 1.

6.2.2.1 Configuration 1

The geometry for this configuration is shown in Figure 6.20. A detail of the position of the antenna is shown in Figure 6.21. The antenna under analysis is one used in the Sport project. It is a horn antenna operating at 85.5 GHz and can be described by its E plane and H plane. For this reason, an RV2 radiation pattern file can be used to describe this antenna. Figure 6.22 represents the E plane and H plane of the Sport antenna. Taking into account the same considerations as for the Jason satellite, the radiation pattern of the on-board antenna was obtained by considering two interactions. Figures 6.23 and 6.24 represent the radiated field for the cuts at $\phi = 0°$ and $90°$.

Figure 6.21 Antenna location in Configuration 1.

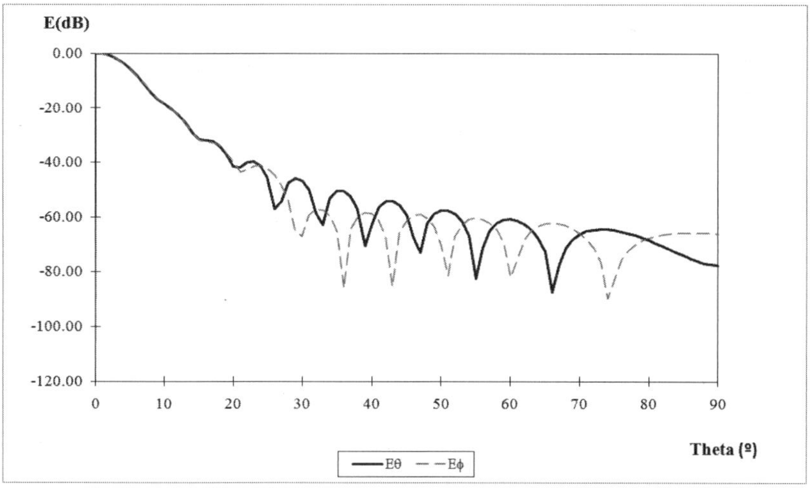

Figure 6.22 Sport project antenna radiation pattern in free space.

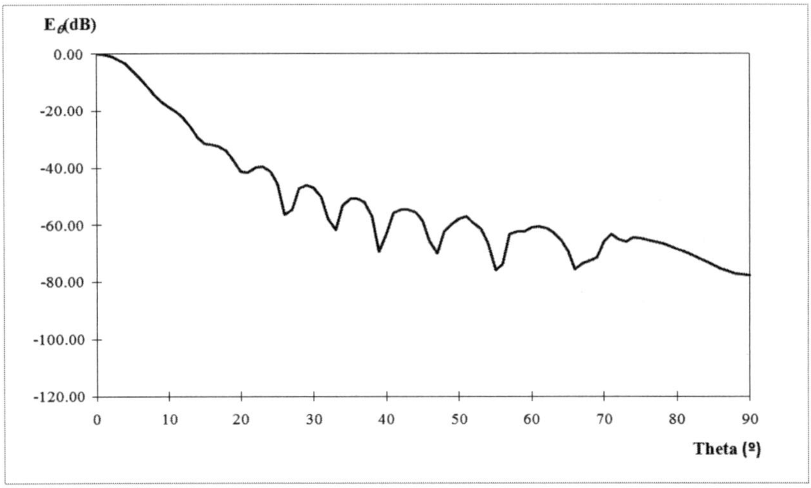

Figure 6.23 θ component of the radiation pattern of the antenna on board the ISS for Configuration 1, $\phi = 0°$.

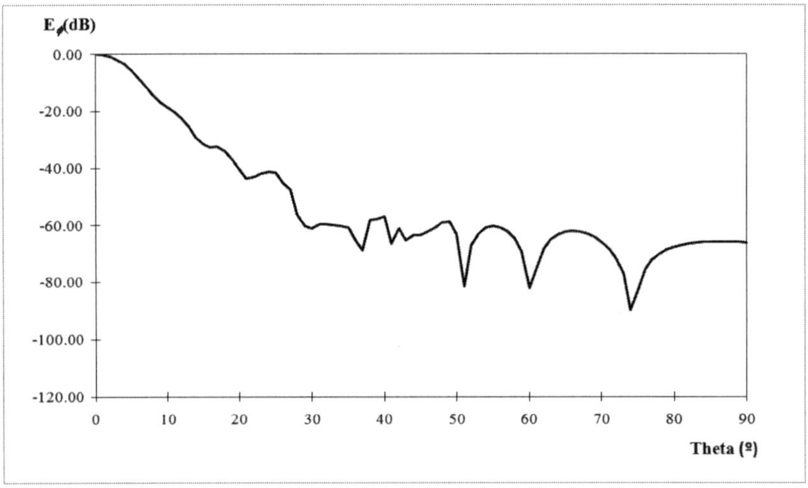

Figure 6.24 ϕ component of the radiation pattern of the antenna on board the ISS for Configuration 1, $\phi = 90°$.

Here it can be observed that the degradation of the radiation pattern was higher for $\phi = 90°$, although it did not affect the main lobe of the pattern.

6.2.2.2 Configuration 2

The geometry of Configuration 2 was obtained by rotating the solar panels 90° (Figure 6.25). This task can be done in any CAGD program. However, the graphical interface of FASANT includes a geometrical feature that allows for rotation of a part of the geometry. Therefore, the modifications of the geometry or the creation of new ones can be performed using FASANT directly, as explained on the DVD. Figures 6.26 and 6.27 show the radiated fields for the same cases as Configuration 1. As in the previous case, the variation in the radiation pattern was not significant.

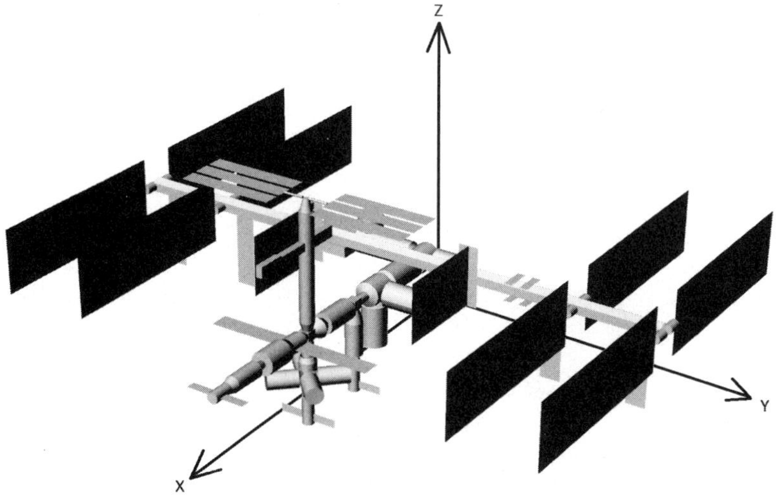

Figure 6.25 ISS configuration 2.

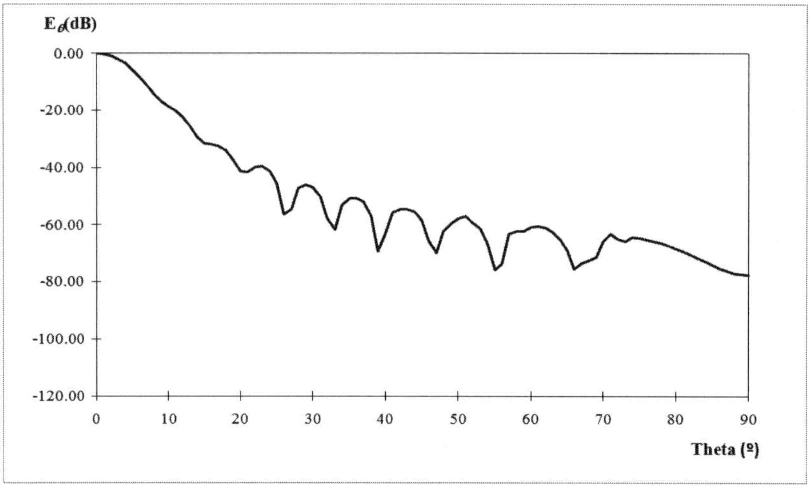

Figure 6.26 θ component of the radiation pattern of the antenna on board the ISS for Configuration 2, $\phi = 0°$.

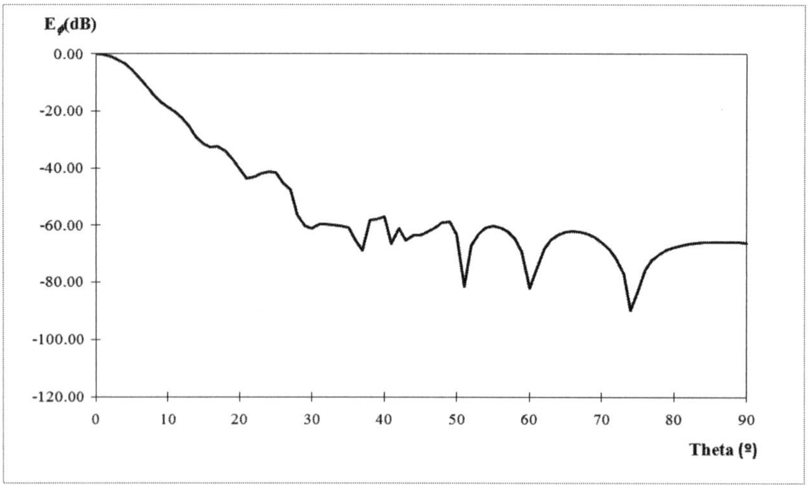

Figure 6.27 ϕ component of the radiation pattern of the antenna on board the ISS for Configuration 2, $\phi = 90°$.

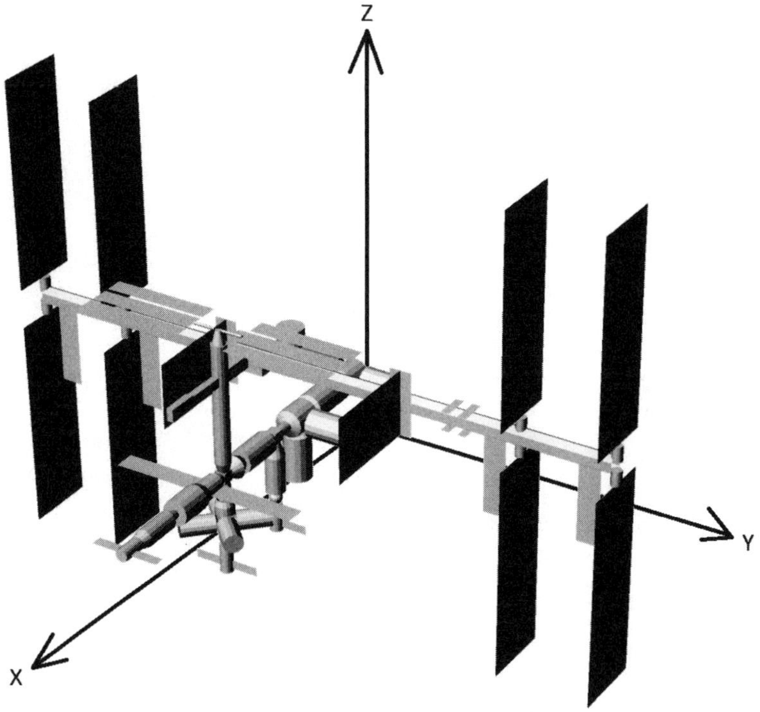

Figure 6.28 ISS Configuration 3.

6.2.2.3 Configuration 3

The geometry for this case is shown in Figure 6.28. This is the worst case because the solar panel can obstruct the direct field radiated by the antenna. Figures 6.29 and 6.30 show the fields radiated for the cuts at $\phi = 0°$ and $90°$. These figures show that for $\phi = 0°$ the radiation pattern was practically the same as that for free space, while at $\phi = 90°$ the panels obstructed the antenna pattern and the side lobes were reduced. However, the main lobe was not affected.

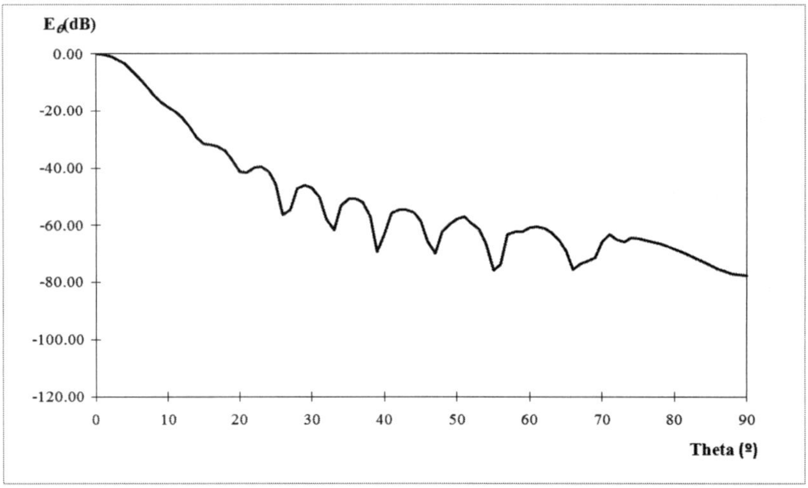

Figure 6.29 θ component of the radiation pattern of the antenna on board the ISS for Configuration 3, $\phi = 0°$.

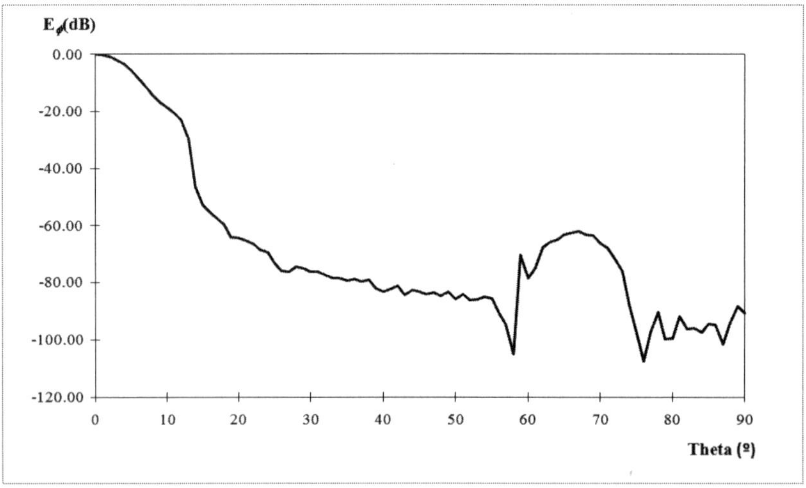

Figure 6.30 ϕ component of the radiation pattern of the antenna on board the ISS for Configuration 3, $\phi = 90°$.

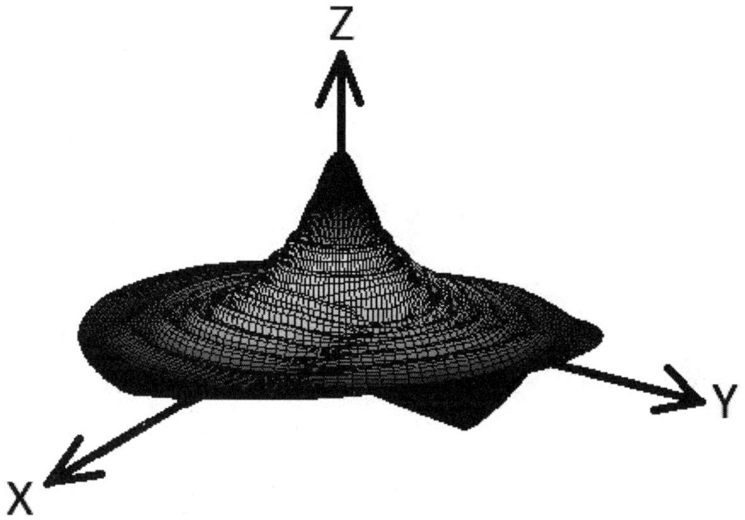

Figure 6.31 Three-dimensional radiation pattern of the Sport project antenna in free space.

6.2.2.4 Three-Dimensional Radiation Pattern

One of the problems in this structure is that the radiation pattern of the on-board antenna could be asymmetric. To analyze this situation, the three-dimensional radiation pattern for the worst case, when the solar panels are close to the main axis of the antenna (configuration 3), was obtained. Figure 6.31 shows this pattern for the antenna in free space. This has been computed from 12 ϕ cuts. Figure 6.32 shows this pattern for the antenna on board the structure. As in the free-space case, 12 ϕ cuts were calculated using FASANT code.

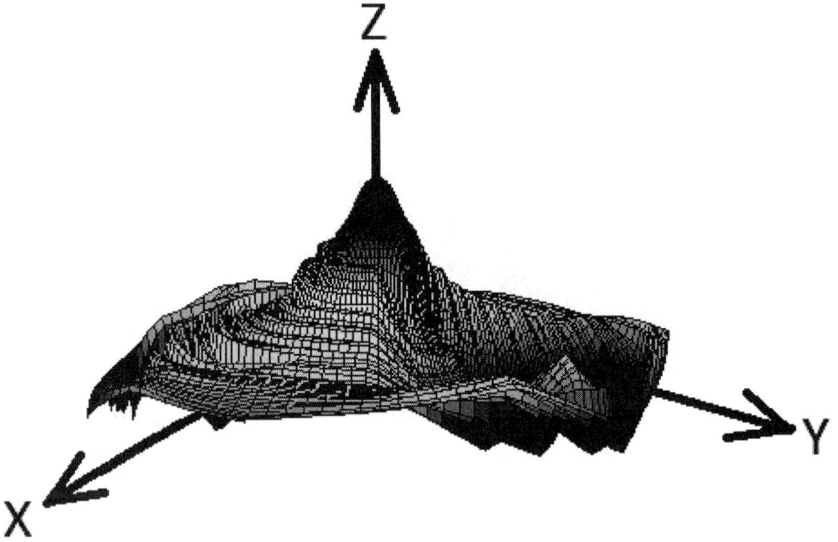

Figure 6.32 Three-dimensional radiation pattern of the Sport project antenna on board the ISS.

6.2.2.5 Analysis of the Cross-Polar Component

The radiation pattern of the input antenna is defined by the E and H planes [17]. Therefore, the radiated field for a given direction of observation is calculated using the following expression:

$$\vec{E} = E_\theta(\theta)\cos\phi\hat{\theta} - E_\phi(\theta)\sin\phi\hat{\phi} \tag{6.3}$$

where $E_\theta(\theta)$ and $E_\phi(\theta)$ are the E-plane and H-plane cuts of the radiation pattern, respectively.

In the ISS analysis, the copolar polarization has been considered in the X direction and the cross-polar polarization has been considered in the Y direction. In spherical coordinates both polarizations are:

$$\hat{P}_C = \cos\phi\hat{\theta} - \sin\phi\hat{\phi} \tag{6.4}$$

$$\hat{P}_X = \sin\phi\hat{\theta} + \cos\phi\hat{\phi} \tag{6.5}$$

The electric field can be expressed in copolar and cross-polar components as the scalar product of the field with $\hat{P}_C$ and $\hat{P}_X$, respectively.

Using expressions (6.4) and (6.5), it is possible to obtain the copolar and cross-polar components of the antenna radiation. When the antenna is on board, degradation in the levels of the cross-polar component can appear. This degradation is due mainly to the diffraction in the solar panels. Figure 6.33 shows the configuration in which this degradation is at a maximum, in which the module containing the antenna is normal to the solar panels. Figure 6.34 shows the copolar and cross-polar components for the antenna in free space. Figure 6.35 shows the results of the copolar and cross-polar components for the antenna on board the ISS for this configuration.

Here, the cross-polar component did not show a substantial increase relative to the antenna in free space. Moreover, these levels were below 45 dB with respect to the values of the main lobe, thus satisfying the antenna specifications.

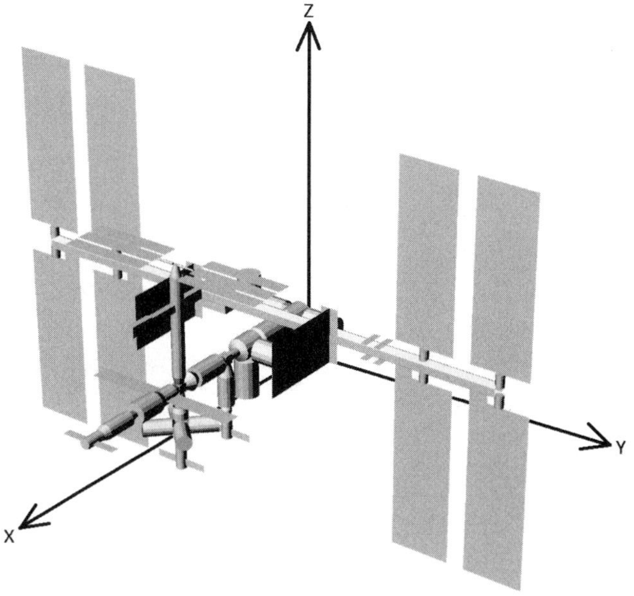

Figure 6.33 Configuration 4.

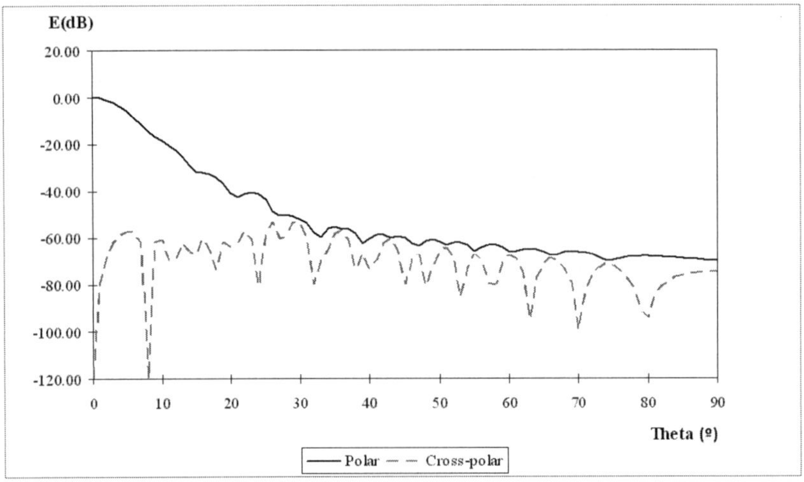

Figure 6.34 Copolar and cross-polar components for antenna in free space, $\phi = 45°$.

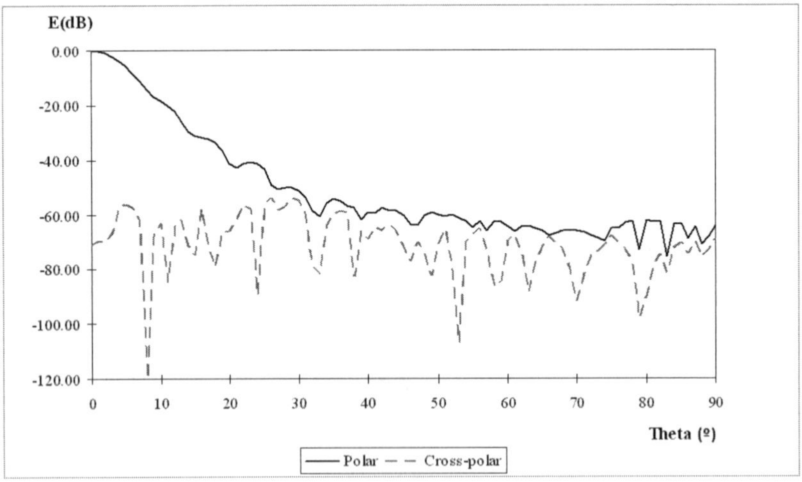

Figure 6.35 Copolar and cross-polar components for the antenna mounted on the ISS, $\phi = 45°$.

6.2.3 Satellite Indoor Reception in Vienna

The new mobile satellite services (MSS), such as satellite digital multimedia broadcast systems or navigation systems, aim to provide coverage to users located in different environments, including inside buildings. For this reason, there is currently great interest in calculating the reception of satellite signals in indoor scenarios. Attenuation of the level of a received signal is mainly caused by reflections and diffractions within the building and those produced in on the nearest external buildings. The GO/UTD method allows for the calculation of these effects by ray-tracing, providing a powerful tool to predict both the power and the channel characteristics of a received signal. Ray-tracing provides all the information necessary to determine parameters such as the power-delay profile, impulse response, and others. The following presents an analysis of satellite-to-indoor coverage using FASANT. This analysis is part of a European Space Agency (ESA) project in which the main objective was to find a computational method to predict the characterization of satellite channels at frequencies close to those assigned to the future Galileo navigation system [23]. The example illustrated in this section consists of a model for the prediction of level propagation inside a building located in the city of Vienna (Figure 6.36) from the reception of the satellite signal used for the Galileo system.

Figure 6.36 Building in Vienna in which the receiver is placed.

A good approximation of the satellite transmission can be achieved by placing the transmitter far from the receiver, for instance, 1 km, to simulate a transmitter located in the far field. Another consideration to take into account is that a satellite has different relative positions with respect to the building because it is geostationary. For this reason, the transmitter was placed in different positions using an igloo configuration, with the location of the receiver antenna at the center of the igloo, as illustrated in Figure 6.37.

Figure 6.37 Transmitter locations.

A geometric model of the building and its surrounding structures (Figure 6.38), typically provided by a CAGD tool, was built from the data obtained in [24]. Usually, the files provided for these tools are created using multiple layers containing a great amount of useless information from a GO/UTD point of view. The first step for creating a DXF file valid for use with FASANT is to eliminate useless layers. For instance, in the example of Figure 6.38, the layer with the vegetation information can be deleted. Another thing to take into account is that these models are usually built for general purposes and they are very detailed. The model description for GO/UTD does not require this level of detail, as only objects larger than the wavelength must be included. With these considerations in mind, the DXF file must be created. One procedure could be to create the facets and surfaces on the original file using a new layer. However, this can be a substantial task for large geometries, as in our example. Another possibility is to export the original model to a DXF format. In this example Rhinoceros was used to edit the geometrical file and, after eliminating the useless layers, the file was exported to a DXF file. Figure 6.39 shows the DXF export options. Choosing 3DFACES for planar surfaces and POLYLINES for curves will generate an exported file that has one of the formats required for FASANT. The final results can be seen in Figure 6.40. As mentioned above, the objective of this example is to determine the indoor coverage of the building represented in Figure 6.36. Figure 6.41 represents the GO/UTD model of this building, whereas Figure 6.42 shows a more detailed view of the indoor room in which the received signal level was obtained.

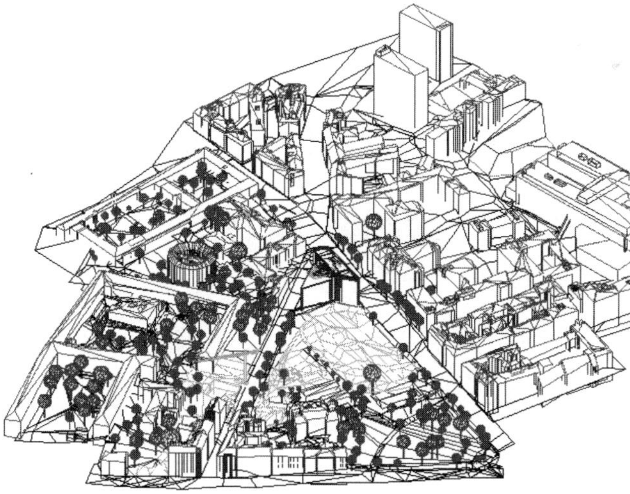

Figure 6.38 General-purpose CAGD model for a part of Vienna.

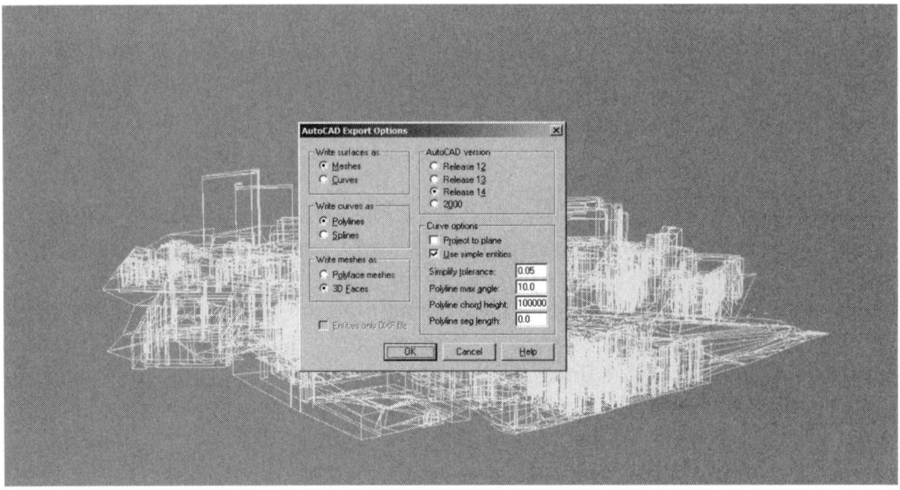

Figure 6.39 DXF export options.

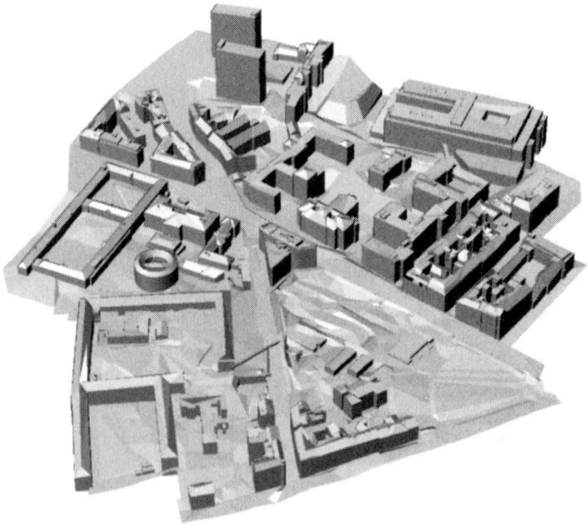

Figure 6.40 GO/UTD model of the same part of the city of Vienna.

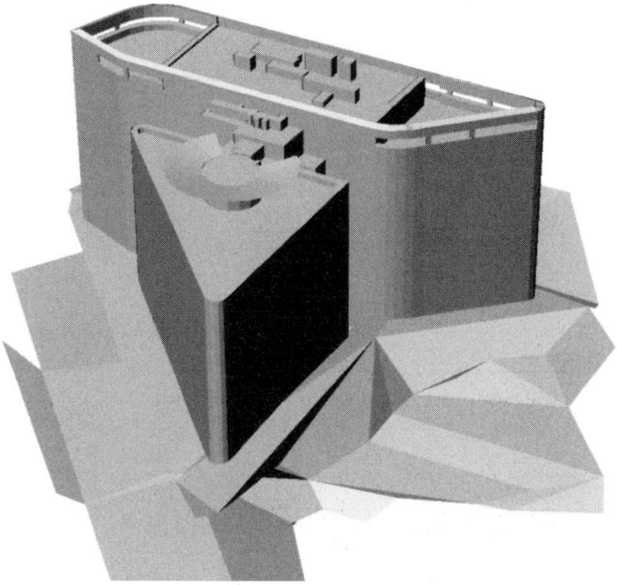

Figure 6.41 GO/UTD office building model.

Figure 6.42 Indoor room detail.

The results were obtained for three different elevation angles with respect to the ground: 30°, 45°, and 60°. Figure 6.43 represents the three different positions with the corresponding ray-tracing; these results have been presented previously [18, 25]

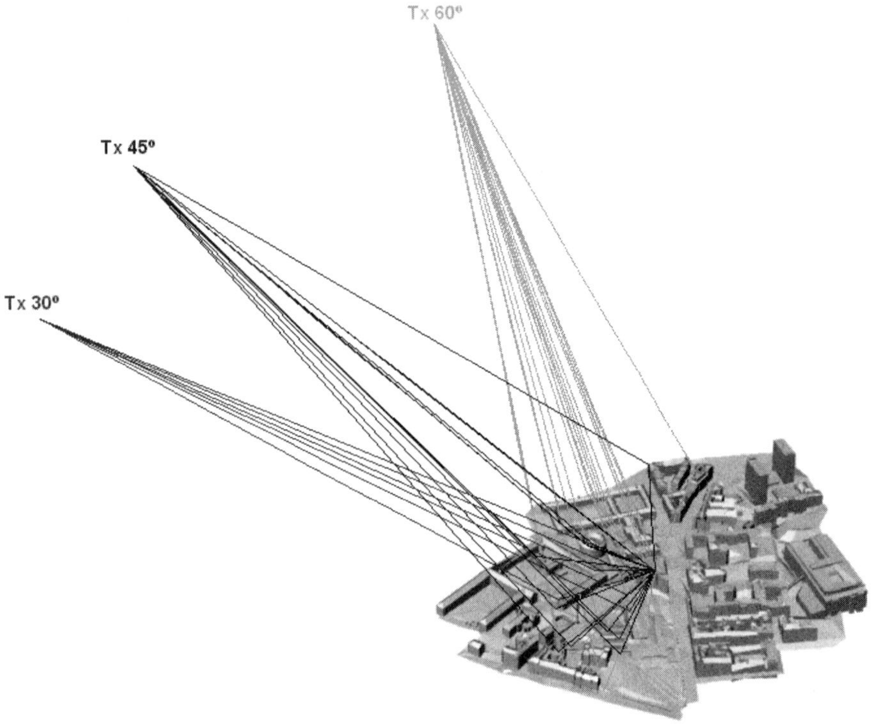

Figure 6.43 Three different transmitter positions for the Vienna case study.

More detailed views of the ray-tracing are shown in Figures 6.44 and 6.45, considering only the building under study (Figure 6.44) and the surrounding part of the city (Figure 6.45).

The transmitter was an omnidirectional antenna in azimuth. As such, the radiation pattern was described by a REV-type antenna file. This means that the antenna can be defined by a single phi cut (see the example on the DVD). The operating frequency was 2.57 GHz and the transmitter power was 40 dBm. In this application, the prediction of the received signal level inside a room should be obtained. Therefore, it is necessary to indicate the position of the receiver. This position was inside the room of Figure 6.42. In this case, four interactions are needed to obtain accurate results, including transmissions through walls, reflections, and diffractions.

Table 6.1 shows the received power in dBm for all cases considering the contribution of the building alone and the contribution of the surrounding area. These values allow an evaluation of the influence of the environment in the final received-power value.

The computational time for this case, using a single Opteron machine at 2.4 GHz, was 34 minutes and 26 seconds for the area of Vienna, modeled with 7,000 facets, considering the 36 Tx positions and four bounces (including diffraction, reflection, and transmission through walls) and 10,000 contributions. Note that the predicted level for 30° and 45° was practically the same for the two cases. In these cases, the main contribution was diffraction by the window of the room closer to the receiver, and the surrounding environment had little influence in the final result. However, for the 60° angle, this diffraction was not so critical, and the reflections from the ground and the surrounding buildings were very important.

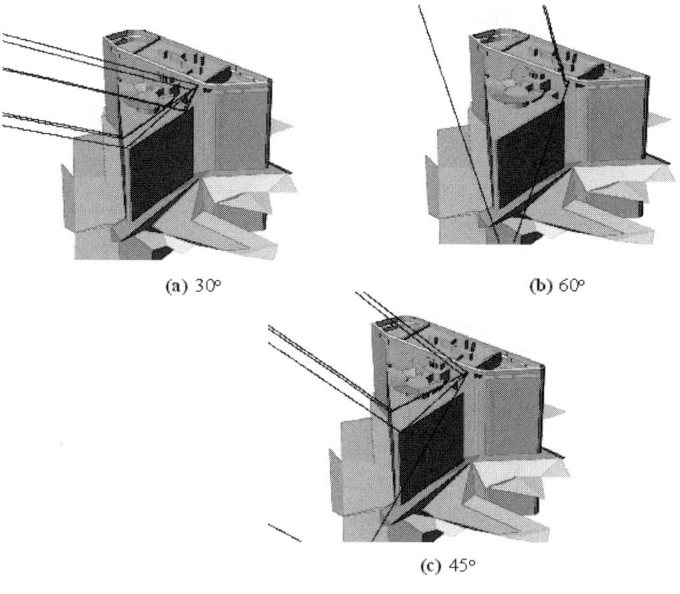

Figure 6.44 Ray-tracing considering the building under study for the different elevation angles.

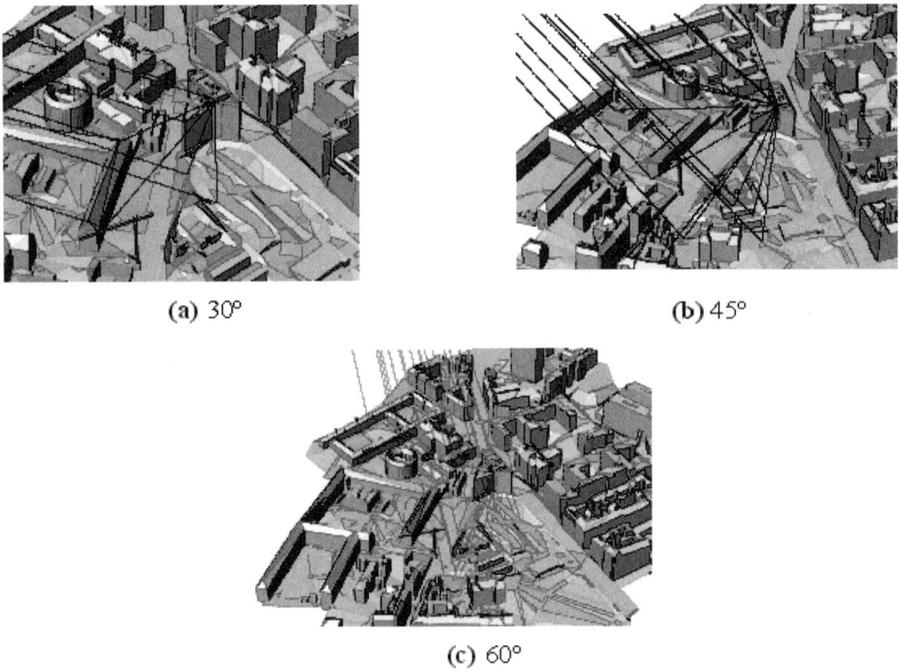

Figure 6.45 Ray-tracing considering the the surrounding enviroment for the different elevation angles.

Table 6.1
Received Power Values for Different Elevation Angles

Tx Location	*Building Alone*	*Vienna*
30°	−57.32	−57.389
45°	−85.60	−84.79
60°	−80.15	−76.72

6.2.4 Wind Turbines

The concern with sustainable energy generation has led to an increase in the construction of wind turbines in recent years. The presence of a high number of these generators poses a serious interference problem for preexisting radio systems in the area due to the obstruction, diffraction, refraction, and scattering of electromagnetic waves caused by the physical structures of the wind turbines. Aside from the higher number of wind generators, it is important to take into account their size. The wind turbines in use today consist of towers exceeding 80m in height and blades exceeding 40m in length. Therefore, they are potentially interfering structures for electromagnetic waves having wavelengths comparable to these sizes.

In this context, the problem presented in this section is the calculation of the scattering field due to a wind generator and the determination of the possible interferences produced by the structure. Because different radio systems could be in the vicinity, the analysis was performed at different frequencies: 100 MHz, 300 MHz, 600 MHz, and 1.2 GHz. The geometric model of the wind generator is shown in Figure 6.46. This model is 120m high and 64m wide; the number of parametric surfaces is 228.

Figure 6.46 Geometrical model of the wind-energy generator.

The first case considered is the scattering field within a radius of 5 km from the wind turbine produced by a dipole placed 20 km away. The reason these distances were chosen is that while wind generators are placed in zones far away from transmission systems, receivers could be closer to them. This case simulates the interferences produced by a wind generator on mobile phone or TV reception. Figure 6.47 represents the case under study.

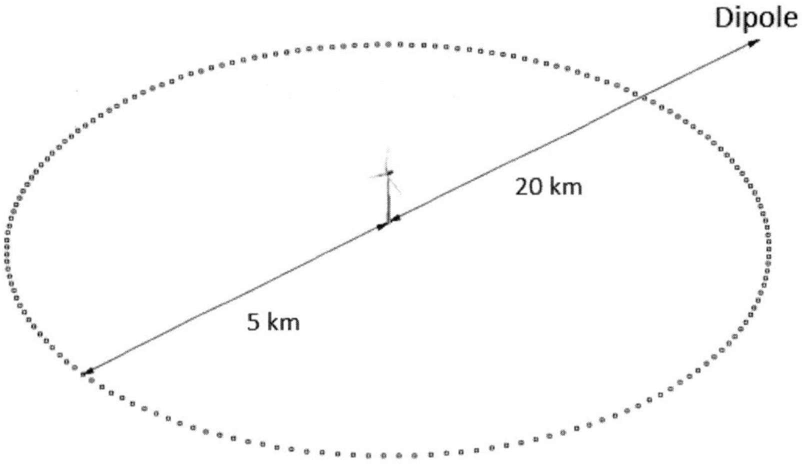

Figure 6.47 Scattered field case for the wind generator.

The interferences here are due only to the field scattered by the wind turbine; therefore, the direct ray was not taken into account. The scattering was computed by considering only the reflected and diffracted rays. The contribution of second- and higher-order interactions was assumed to be negligible.

The results obtained with GO/UTD were compared with the MoM (with MONURBS code) and PO by evaluating the integral numerically, as explained in Section 4.3.1. The frequencies analyzed were 100 MHz (Figure 6.48), 300 MHz (Figure 6.49), 600 MHz (Figure 6.50), and 1.2 GHz (Figure 6.51). These results have been presented previously [18].

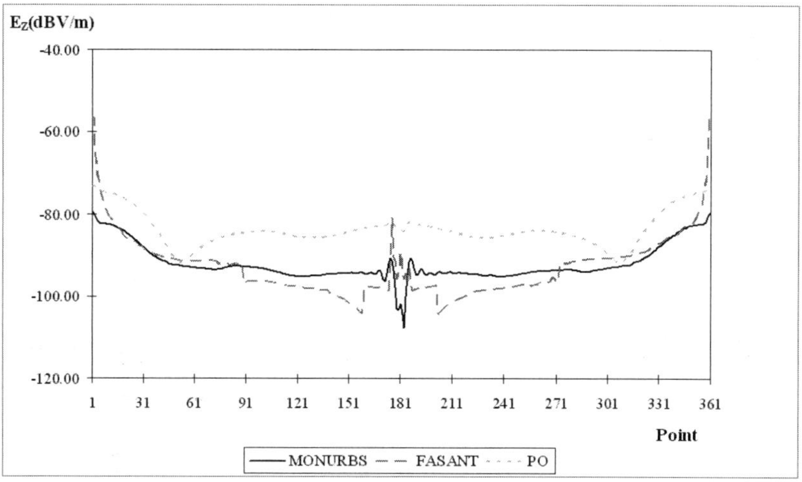

Figure 6.48 Scattered field at 100 MHz, comparing FASANT results with MoM and numerical PO.

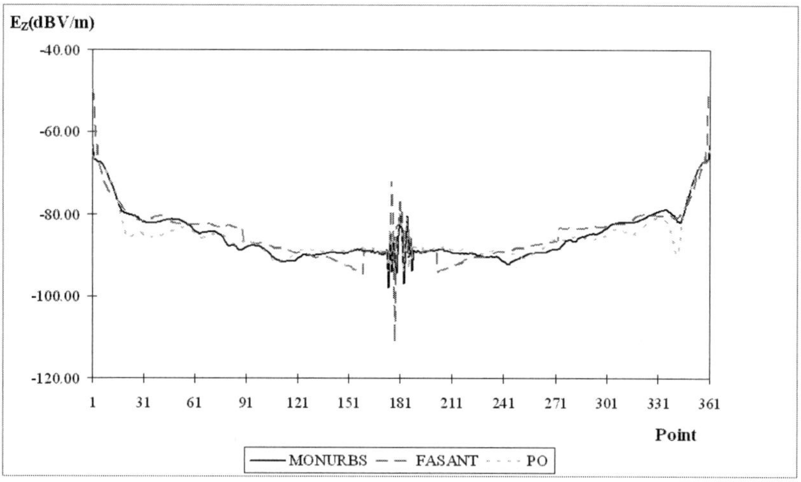

Figure 6.49 Scattered field at 300 MHz, comparing FASANT results with MoM and numerical PO.

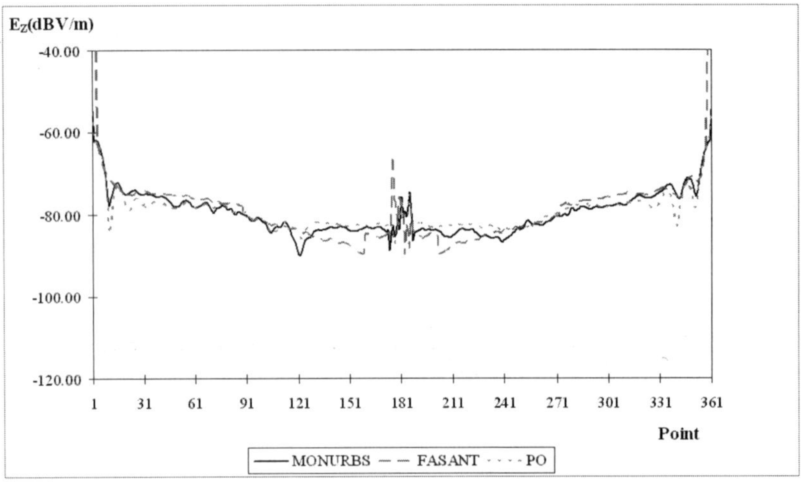

Figure 6.50 Scattered field at 600 MHz, comparing FASANT results with MoM and numerical PO.

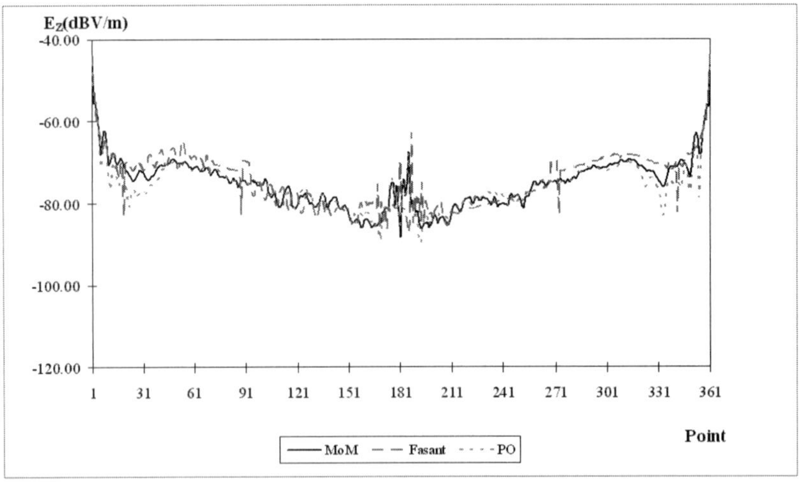

Figure 6.51 Scattered field at 1.2 GHz, comparing FASANT results with MoM and numerical PO.

Figure 6.48 shows that the GO/UTD results were significantly different from the MoM results. This is because the asymptotic approximation was not good enough at 100 MHz. The wavelength in this case was 3m, and there are parts much lower than this size. For higher frequencies, the approximation was good enough and the predicted levels were comparable with the MoM method.

The advantage of GO/UTD is, obviously, the computational time. It took 2 hours and 30 minutes for all the frequencies using a single-processor PC to analyze the wind turbine from 100 MHz to 20 GHz (10 frequencies). The principal task was the ray-tracing computation, and this is independent of the frequency. However, the MoM must calculate the induced currents for each frequency, and, therefore, the computational time increases with the frequency. For instance, the 1.2-GHz case took 2 hours of run time using the MONURBS computer code with a 12-processor computer. Figure 6.52 shows the scattering field for 2.4 GHz, 4.8 GHz, 9.2 GHz, and 19.2 GHz; these results were computed in the same simulation as the previous one. There are not MoM method results for these frequencies because the memory requirements were beyond the capacity of the computer used.

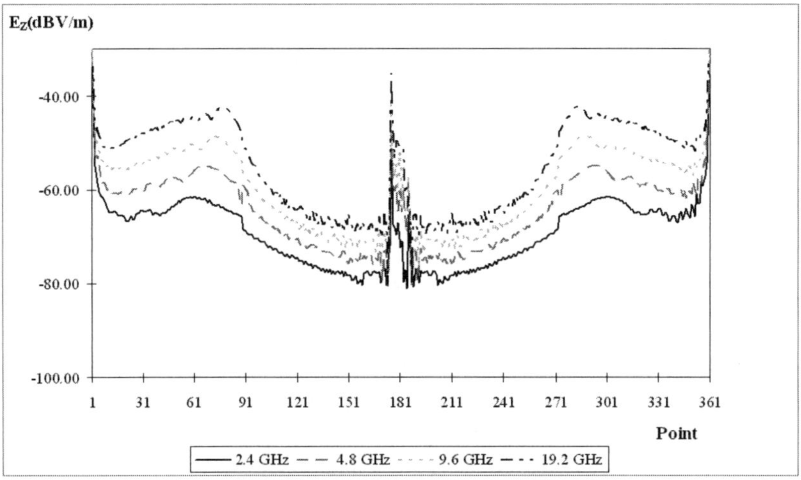

Figure 6.52 Scattered field at 2.4 GHz, 4.8 GHz, 9.6 GHz, and 19.2 GHz computed with FASANT.

Another case studied consisted of placing the transmitter and the receiver in the same place (monostatic scattering). Only one frequency was considered here, because in this case the MoM requires a higher computational time as the induced currents must be recalculated for every transmitter position, while for the

GO/UTD, the computational time was the same that the previous case. The results are presented in Figure 6.53.

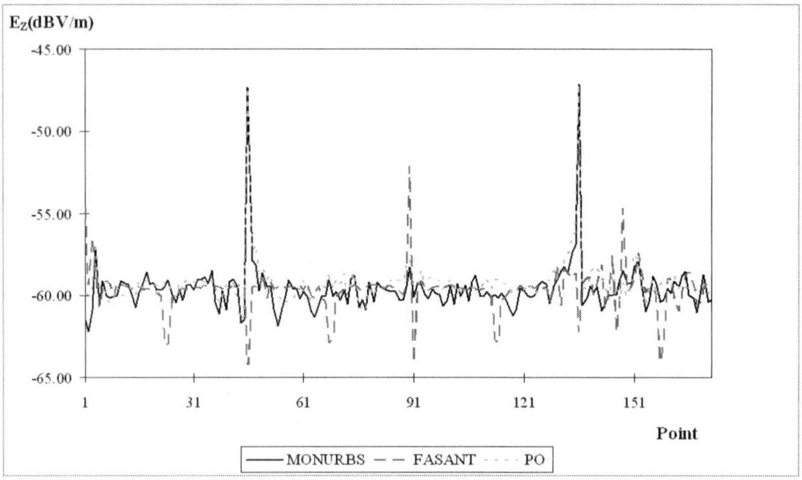

Figure 6.53 Monostatic scattering.

6.3 RESULTS OBTAINED WITH POGCROS AND NUMERICAL EVALUATION OF THE PO INTEGRAL

POGCROS is a computer tool for analyzing the monostatic and/or bistatic radar cross-sections (RCS) of large, electrically complex targets. This computer tool is based on PO/PTD combined with GO to compute the multiple interactions, as was presented in Chapter 4. Perfectly electrically conductive (PEC) surfaces can be analyzed as well as surfaces composed of dielectric materials or radar-absorbing material (RAM). The tool includes the acceleration techniques presented in Chapter 5 to speed up the computation of the RCS.

POGCROS can be used in such areas as ballistic missile-defense applications, automatic target recognition (ATR) data calculation, inverse synthetic aperture radar (ISAR) image generation, and low-observable (stealth) object design, among others. The following sections present three RCS and ISAR examples analyzed using POGCROS. The first is an example of the monostatic RCS of a Scud missile, the second corresponds to the range profiles of a ship, and the third is an example of a cavity analysis.

6.3.1 RCS of a Scud Missile

Scud is a series of tactical ballistic missiles developed by the Soviet Union during the Cold War and widely exported to other countries. The term comes from the NATO reporting name SS-1 Scud which was attached to the missile by Western intelligence agencies. The Russian names for the missile are the R-11 (the first version), R-17, and R-300 Elbrus (later developments). The name Scud has been widely used to refer to these missiles and the wide variety of derivative variants developed in other countries based on the Soviet design. Figure 6.54 shows a picture of a Scud missile.

Figure 6.54 Scud missile.

In this example, the Scud monostatic RCS was computed using POGCROS code. To validate the asymptotic techniques presented in this book, these results were compared with MoM. This model is 11m long and 3m wide. The number of parametric surfaces in the model is 57. The analysis was carried out at 1 GHz.

The first step is to create the geometric model. As discussed in the case of GO/UTD, this task can be performed starting from the missile planes using multiple CAGD programs like Rhinoceros, Microstation, Catia, and AUTOCAD, among others. Once the geometry has been created, it must be exported to a file containing the necessary geometrical data to apply the PO method. POGCROS code can input IGES, FC, SAT, and DXF formats. IGES format is a standard, and it is available in most of the CAGD programs. FC format is a simplified IGES format that uses only the IGES's 128 primitive, which describes the NURBS surfaces.

The geometry introduced to POGCROS was created based on missile dimensions. As mentioned in previous chapters, POGCROS is based on PO/PTD. Therefore, when making the geometric model sizes lower than the size of the wavelength should not be modeled because PO/PTD restrictions. The geometric model used for the computation is shown in Figure 6.55. Once the geometry has been defined, the parameters for RCS calculations must be fixed.

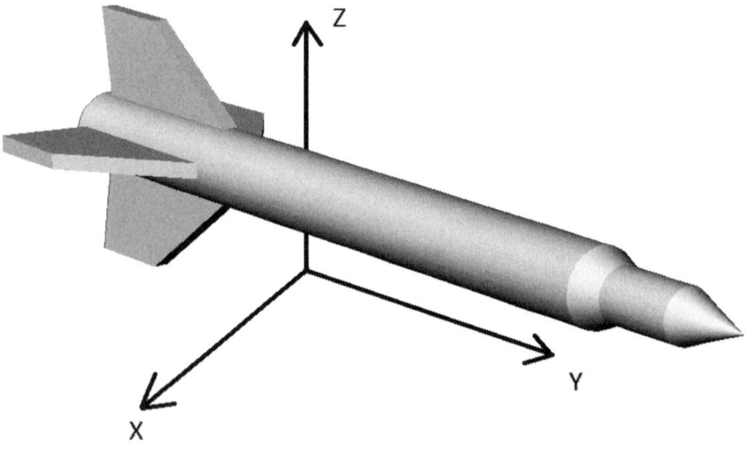

Figure 6.55 CAGD model of the Scud missile.

The parameters are:

- Polarization of the incident wave: two polarizations, vertical and horizontal, were considered for each case.
- Type of simulation: in this example a monostatic analysis of RCS has been chosen.
- Sweep in frequency: in this example, the RCS will be analyzed at 1 GHz.
- Angular sweep type: other important parameters are the number of samples in the angular sweep and the number of cuts. In this example, three different cuts were analyzed: a cut at $\theta = 90°$ and two cuts at $\phi = 0°$ and $90°$, respectively, were chosen.
- Interactions: as in GO/UTD, the order must be selected depending on the geometric model and the angular sweep that is analyzed. In this case, a second order was selected due to the presence of dihedrals on the geometry.

Figures 6.56 to 6.58 show comparisons of the RCS computed with POGCROS and with MoM (MONURBS code) for vertical polarization, while Figures 6.59 to 6.61 show the results for horizontal polarization.

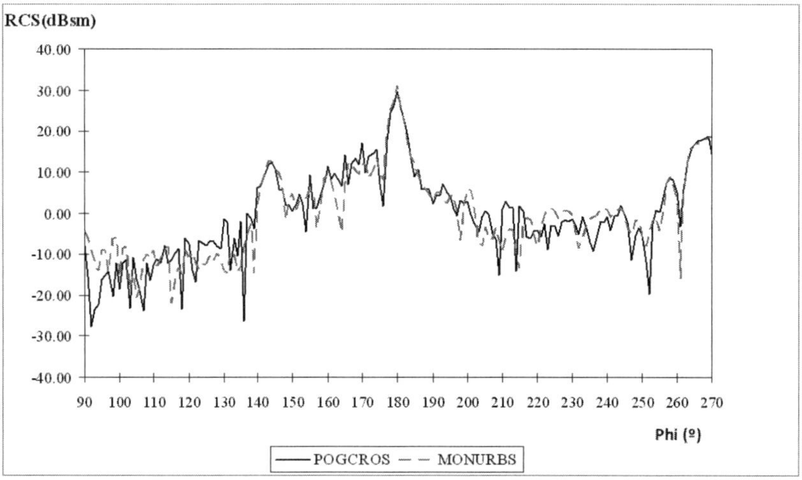

Figure 6.56 RCS comparison between POGCROS and MONURBS for the cut at $\theta = 90°$ and vertical polarization.

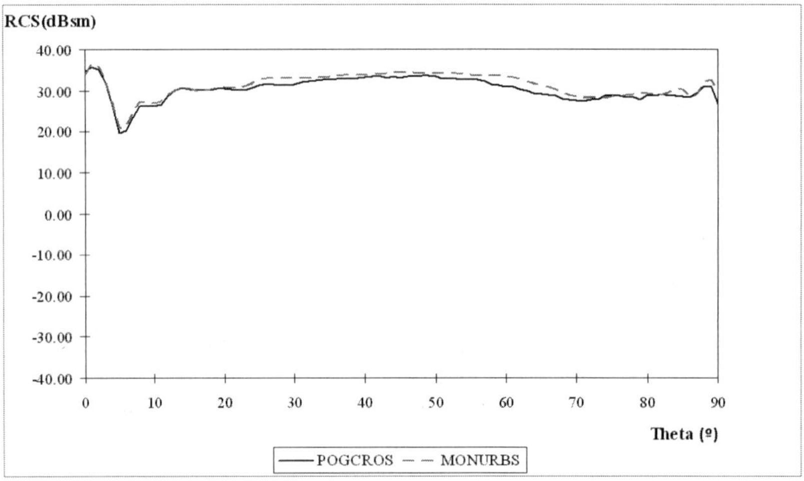

Figure 6.57 RCS comparison between POGCROS and MONURBS for the cut at $\phi = 0°$ and vertical polarization.

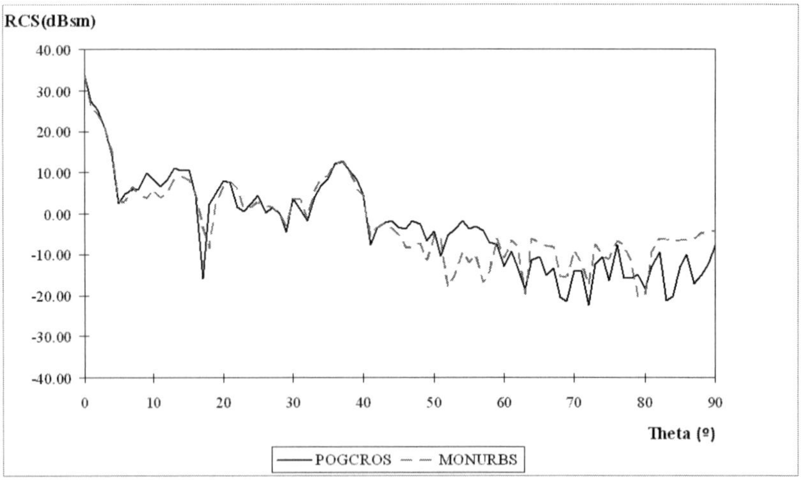

Figure 6.58 RCS comparison between POGCROS and MONURBS for the cut at $\phi = 90°$ and vertical polarization.

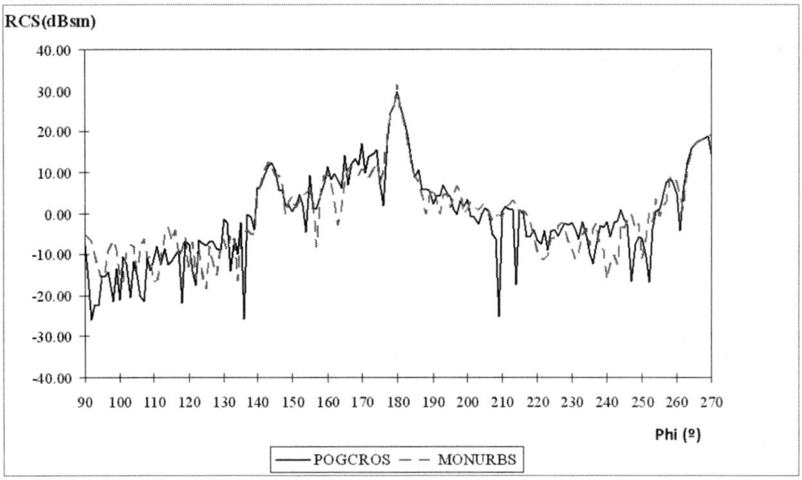

Figure 6.59 RCS comparison between POGCROS and MONURBS for the cut at $\theta = 90°$ and horizontal polarization.

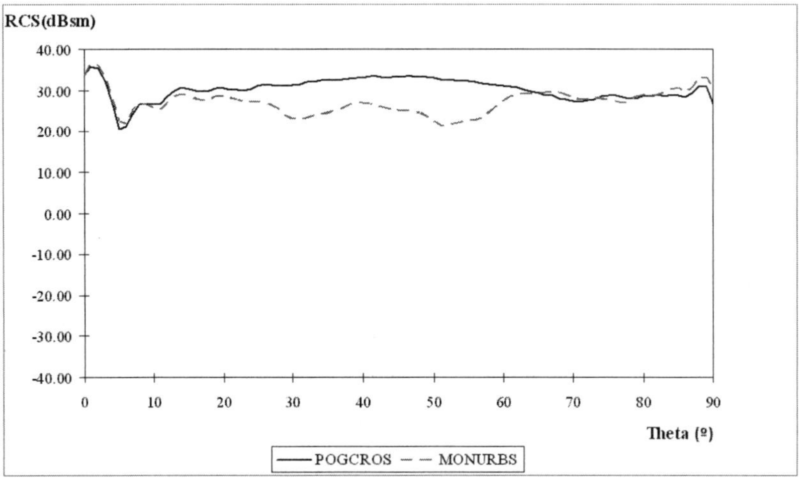

Figure 6.60 RCS comparison between POGCROS and MONURBS for the cut at $\phi = 0°$ and horizontal polarization.

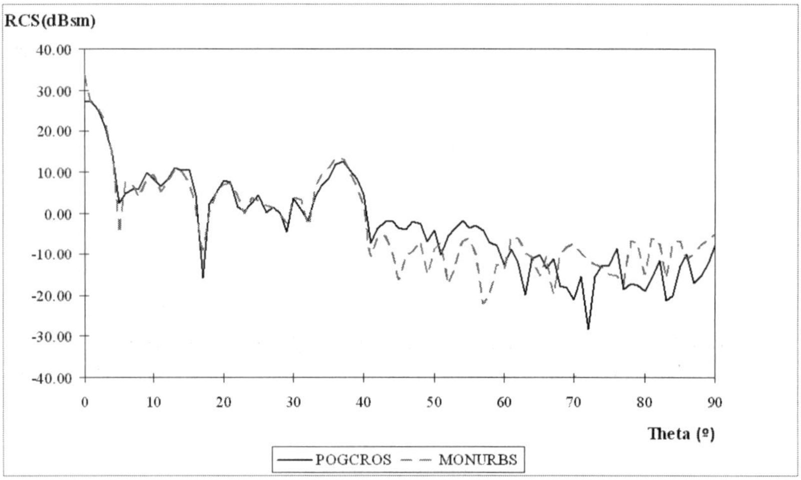

Figure 6.61 RCS comparison between POGCROS and MONURBS for the cut at $\phi = 90°$ and horizontal polarization.

All the results shown above demonstrate the good agreement between a rigorous method like MoM and the asymptotic techniques presented in this book. However, the computational time is quite different for the two methods. The CPU time needed to obtain the values for the 181 directions with POGCROS was 5 minutes, 41 seconds on a PC with an INTEL Core 2 Duo (only one processor was used) at 1.87 GHz, and with MoM it was 1 hour, 50 minutes, 15 seconds on a Blade 24-GB Intel Nehalem 2.0 GHz eight-core CPU (eight processors were used), showing the capability of the asymptotic methods for obtaining accurate results on complex bodies with a low computational cost.

The predicted levels were practically identical for both methods because the asymptotic approximation was sufficient for 1.2 GHz.

6.3.2 Analysis of the Range-Profile Image of the Tabarca Ship

Knowledge of the RCS of different vessels is a critical aspect in surface radars used for coastal surveillance. The identification of individual vessels using their range profiles can be done if their theoretical RCS are known. In this section, a study of the range profile of the Tabarca ship from the RCS results obtained from POGCROS was performed. Tabarca is a Spanish Navy patrol boat dedicated to coastal watch.

In this example, the objective was to compute the range profile image for the Tabarca ship. To achieve this, the monostatic RCS in several samples, both in angular sweep and frequency, were computed using POGCROS code. The first step was to create the geometric model from the ship planes using CAGD programs and export these models to an IGES or FC format.

The geometric model used for the computation is shown in Figure 6.62. It has a length of 44.26m and a beam of 6.6m and is made of steel. For the application of an asymptotic technique, only the parts of the ship that are large with respect to the wavelength were modeled. The many cables of the ship and other details smaller than the wavelength were not modeled. Even when the cables are constructed of steel, their contribution to the total RCS is negligible.

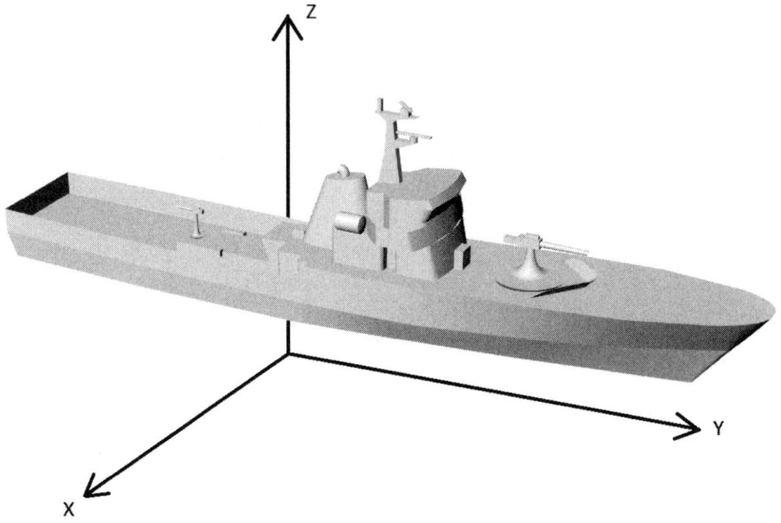

Figure 6.62 Geometric model of the Tabarca ship.

As seen in Figure 6.63, which shows details of the boat, all cables and elements with negligible influence to the RCS were removed from the model.

Figure 6.63 Detailed view of Tabarca.

Once the geometry is defined, the computation parameters must be fixed. In this example, as mentioned above, the monostatic RCS at several angles and frequencies was obtained. A second order of interactions was selected for the same reasons as in the previous cases. Once the computations were performed, post-processing of the results was necessary to compute the range profiles, which were used to determine the scattering centers of the body (i.e., the areas in which this body contributes more significantly to the scattered field).

Range profile information was obtained by illuminating the target within a frequency bandwidth. The phase change in the echoes of the scattering centers required for computing the range profile position were obtained by sweeping the frequency. With this information, the physical scattering locations can be reconstructed via the Fourier transform. The main limitation of this procedure is the resolution, which increases with the bandwidth according to the following expression [26]:

$$A_r = \frac{c}{2\Delta f} = \frac{\lambda}{2}\frac{1}{(\Delta f/f)} \tag{6.6}$$

The range profile resolution is thus increased by increasing the frequency band over which the model is analyzed.

The steps required to obtain a range profile are:

1. Specify the approximate resolution desired.
2. Compute the body-scattered field at several frequencies; pad the data array with zeros to obtain an array (linear or square) that is a power of two in length and perform the FFT.
3. Resort the data and display the 1-D, 2-D, or 3-D image.

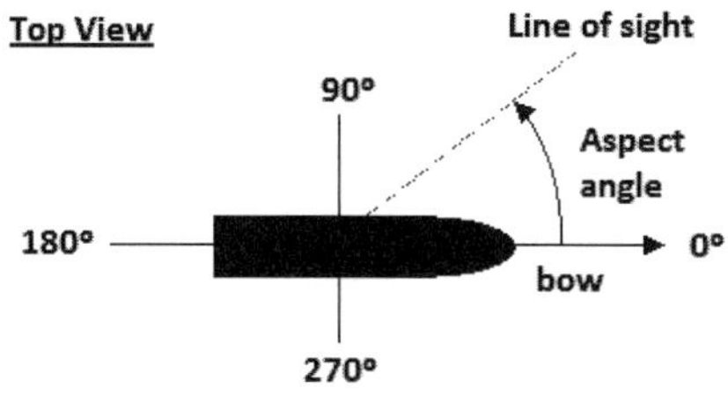

Figure 6.64 Top view of Tabarca with the aspect angle used for the computation of the range profile.

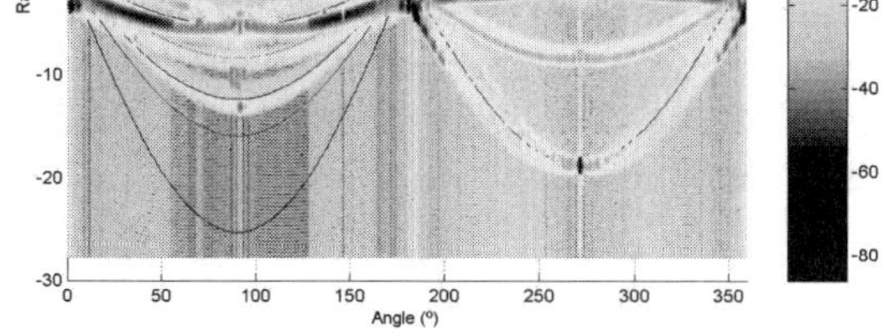

Figure 6.65 Range profile versus azimuth angle for the Tabarca ship in the harbor obtained from the POGCROS results; the polarization is HH.

In these experiments, the Tabarca RCS was measured from 10 to 10.37 GHz. At each frequency, the incidence angle was changed to cover a $360°$ sweep of the ship for a cut at $\theta = 90°$ (see Figure 6.64).

Figure 6.65 shows the range profile obtained from the RCS computation of POGCROS. These results were compared with the measurements, shown in Figure 6.66.

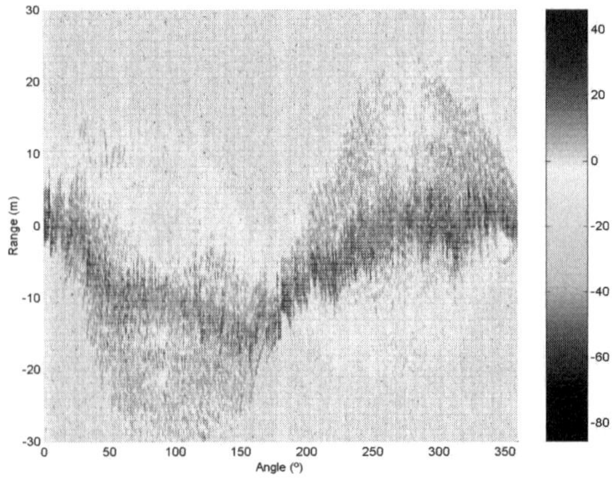

Figure 6.66 Range profile versus azimuth angle for the Tabarca ship in the harbor obtained from measurements; the polarization is HH.

These figures show relatively good agreement between the calculated and measured results in terms of shape and magnitude although the CAGD model of the ship is strongly simplified relative to the real ship. The range profile from the measurements, however, shows some ripples that do not appear in the computations. There are two main reasons for this: first, as the ship is in the harbor, its position is not stable and its rotational center seems to wobble, and second, the real ship has more scattering centers than the CAGD model, which cannot include all the details. These scattering centers can also cause the ripples in the range profile map.

6.3.3 Cavity Analysis

The RCS analysis of open cavities is an important issue in bodies with large ducts because these can contribute significantly to the total RCS. For electrically large cavities, asymptotic methods are computationally more efficient than rigorous ones. The problem with the asymptotic techniques is the large number of bounces that must be considered in the analysis to obtain accurate results. POGCROS can calculate a large number of interactions and can therefore provide a good analysis for this kind of problem.

In this section, an example of an open cavity was analyzed using POGCROS code and validated with the MoM technique. A geometric model was created with a CAGD computer tool, and it was exported to an IGES or FC format. This model has 185 planar surfaces, and its dimensions are 54 cm long, 27 cm wide, and 21 cm high. The geometric model is shown in Figure 6.67.

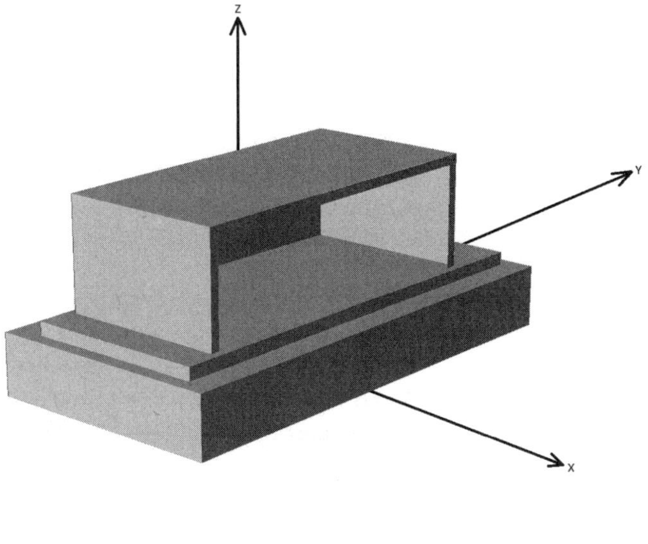

Figure 6.67 Geometric model of the cavity.

The next step was to select the parameters required to perform the simulation. Vertical polarization was analyzed. The RCS was computed for a cut at $\theta = 90^\circ$ and ϕ varying from 0° to 90°. The analyzed frequency was 35 GHz. Due to analyzing an open cavity, multiple interactions will exist inside. Therefore, in this case a fifth order was selected.

The comparison with the MoM technique is shown in Figure 6.68. These results have been presented previously [27].

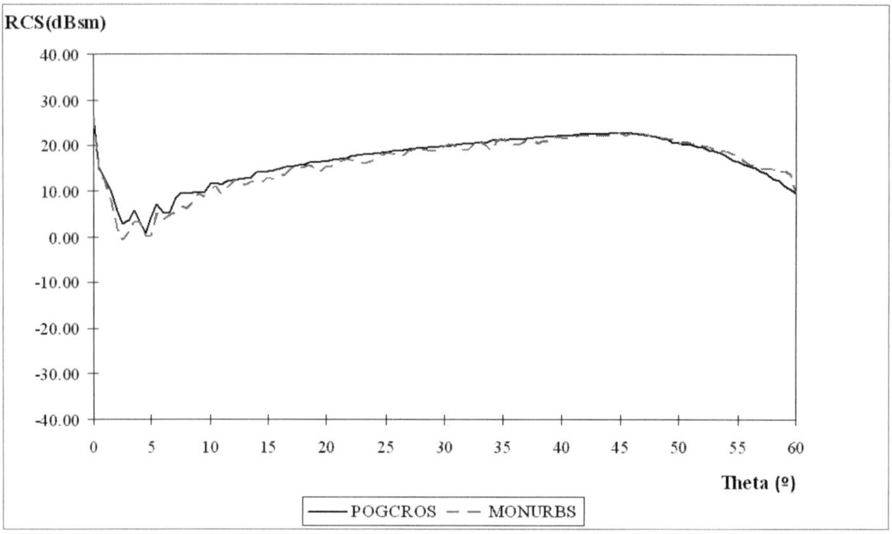

Figure 6.68 RCS comparison between POGCROS and MONURBS for a cut $\phi = 90°$ and vertical polarization.

A good agreement between both results was observed. The advantage of the asymptotic analysis is here, as in the previous examples, the computational cost. The CPU time required to obtain the results with POGCROS was 10 minutes, 36 seconds on a PC with an INTEL Core 2 Duo (only one processor was used) at 1.87 GHz, and with MoM it was 8 hours and 35 minutes using an AMD Opteron with four processors at 2.4 GHz.

6.3.4 Analysis of a Reflector Antenna

The analysis and design of reflector antennas are a common problem due to the extensive use of such antennas. Due to the large size of the reflectors, the asymptotic techniques are very useful tools. For parabolic reflectors the antenna is placed in the focus. This is a problem when ray-tracing techniques are applied, because a caustic appears and the predicted field diverges. For this reason GO/UTD cannot be applied to this problem and PO becomes the optimal method for the analysis.

In this section a reflector was analyzed using PO, evaluating the integral numerically as shown in Section 4.2.1. The results were compared with the MoM technique.

The reflector is an offset parabolic with diameter of 27 cm. The feed is a circular horn with a diameter of 7.5 cm. The frequency of analysis was 16 GHz. Figure 6.69 shows the reflector with the feed (horn antenna in the focus).

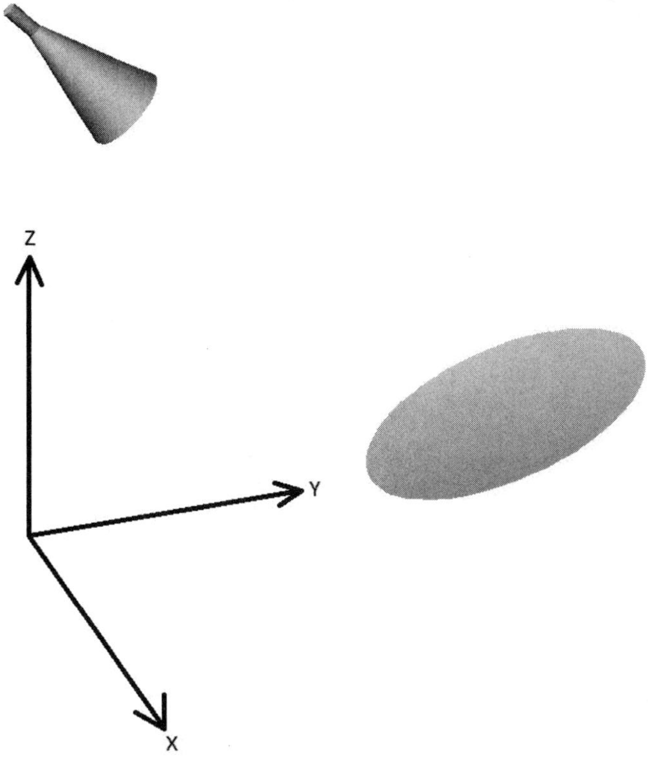

Figure 6.69 Horn antenna and parabolic reflector.

The scattering was computed by solving the PO integral numerically, as explained in Section 4.2.1. The results obtained were compared with the MoM technique. Figures 6.70 to 6.75 show comparisons between the PO and MoM results for cuts at $\phi = 0°$, $45°$, and $90°$ and the θ and ϕ components.

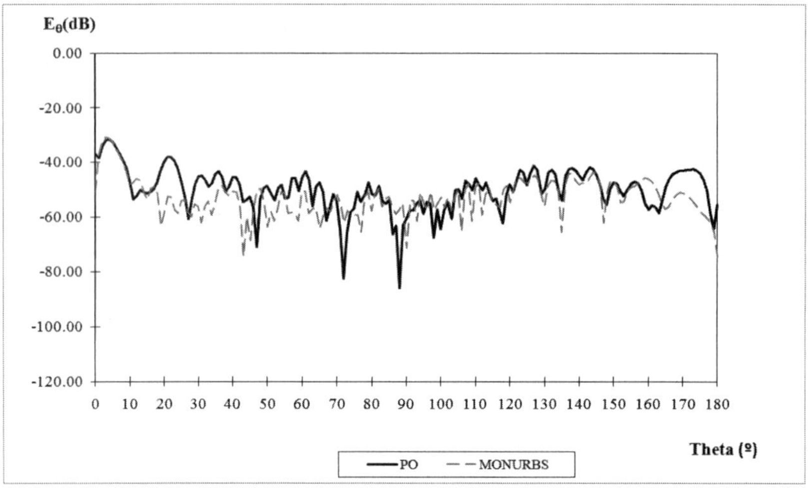

Figure 6.70 PO and MONURBS comparison for a cut with $\phi = 0°$ and the θ component.

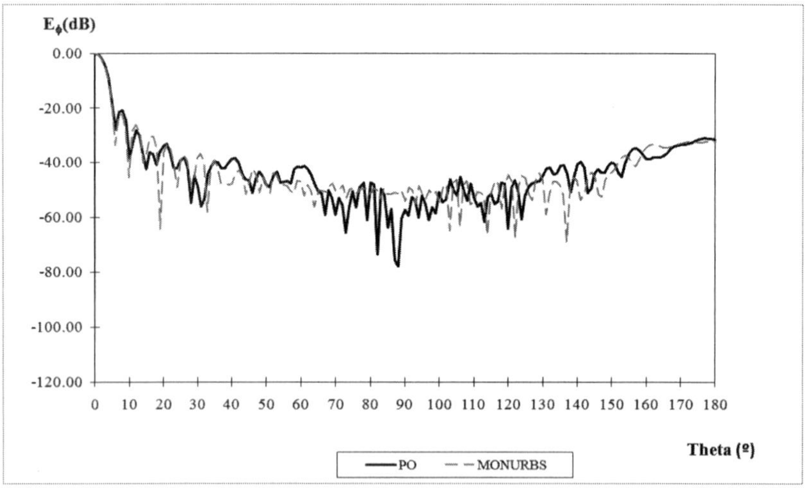

Figure 6.71 PO and MONURBS comparison for a cut with $\phi = 0°$ and the ϕ component.

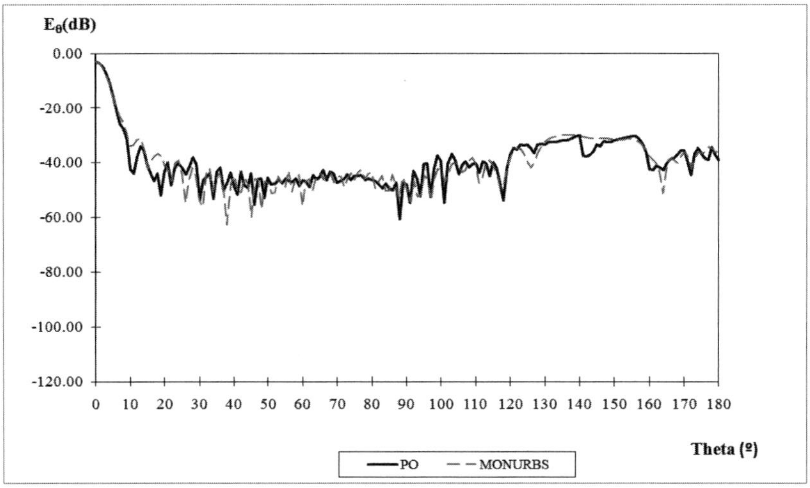

Figure 6.72 PO and MONURBS comparison for a cut with $\phi = 45^\circ$ and the θ component.

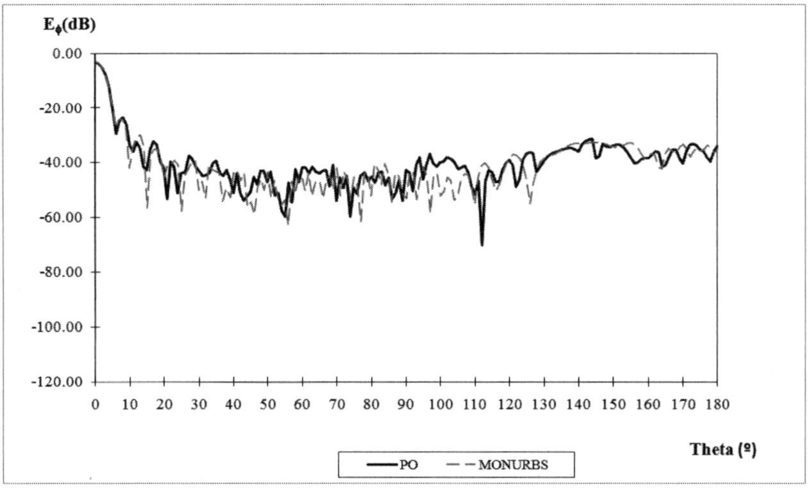

Figure 6.73 PO and MONURBS comparison for a cut with $\phi = 45^\circ$ and the ϕ component.

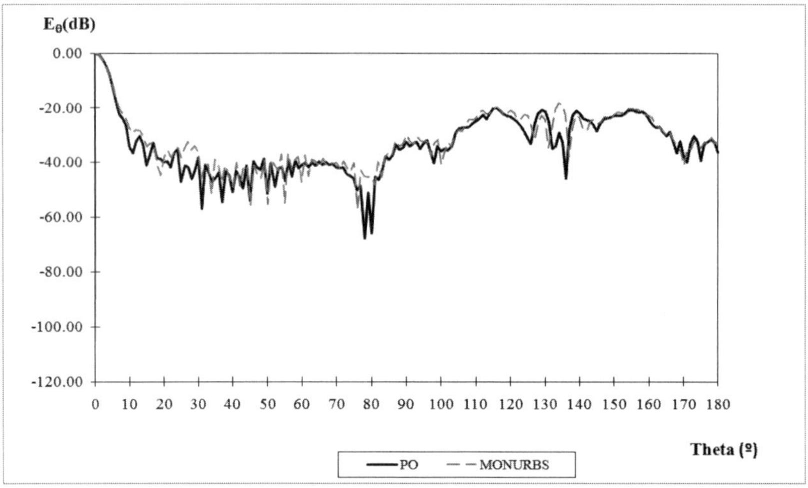

Figure 6.74 PO and MONURBS comparison for a cut with $\phi = 90°$ and the θ component.

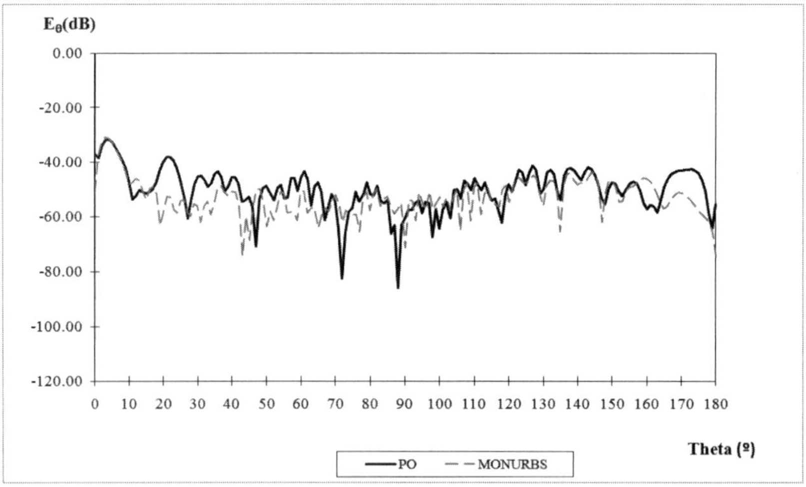

Figure 6.75 PO and MONURBS comparison for a cut with $\phi = 90°$ and the ϕ component.

Good agreement was seen between the two methods except for the last case, in which the field level was considerably below the maximum of the radiated field, and, therefore, was at the noise level. However, the computational times were different. The results from both methods were obtained using the same computer: an Intel Xeon with a dual core at 3 GHz. The computational time for the numerical PO was about one minute using a single processor while for MoM it was 4 minutes using four processors. As in the previous cases, this proves that asymptotic methods obtain accurate results with a lower computational cost.

References

[1] J. Pérez, F. Saez de Adana, O. Gutiérrez, I. González, M. F. Cátedra, I. Montiel, and J. Guzmán. "FASANT: Fast computer tool for the analysis of on board antennas." *IEEE Antennas and Propagation Magazine*, 41(2):94–98, April 1999.

[2] L. Lozano, M. J. Algar, I. González, and M. F. Cátedra. "FASANT: a versatile tool to analyze antennas and propagation in complex environments." *3rd European Conference on Antennas and Propagation 2009*, pp. 2088–2092, 2009.

[3] M. Domingo, F. Rivas, J. Pérez, R. P. Torres, and M. F. Cátedra. "Computation of the RCS of complex bodies modeled using NURBS surfaces." *IEEE Antennas and Propagation Magazine*, 37(6):36–47, December 1995.

[4] L. Lozano, M. J. Algar, M. Blanco, I. González, and M. F. Cátedra. "Computer code for the efficient calculation of the RCS of complex target considering multiple bounces." *IEEE Antennas and Propagation International Symposium*, 2008.

[5] Web page of FEKO: www.feko.info

[6] F. Molinet and C. Louet. "SARGASSES: Interactive computer program with graphics based on GTD, for antennas mounted on an aircraft." *Antennas and Propagation International Symposium*, pp. 633–636, 1986.

[7] Web page of REMCOM: www.remcon.com

[8] M. F. Cátedra and J. Pérez Arriaga. *Cell Planning for Wireless Communications*. Artech House, 1999.

[9] Web page of SUPERNEC: www.supernec.com

[10] M. Sabbadini, G. Guida, and M. Bandinelli. "The antenna design framework-electromagnetic satellite." *IEEE Antennas and Propagation Magazine*, 51(2):225–235, April 2009.

[11] Web page of TICRA: www.ticra.com

[12] D. Andersh, J. Moore, S. Kosanovich, D. Kapp, R. Bhalla, R. Kipp, T. Courtney, A. Nolan, F. German, J. Cook, and J. Hughes. "Xpatch 4: The next generation in high frequency electromagnetic modeling and simulation software." *Radar Conference*, pp. 844–849, 2000.

[13] Web page of SAIC: www.saic.com

[14] R. F. Harrington. *Field Computation by Moment Methods*. MacMillan, 1968.

[15] W. C. Chew, J. Jin, E. Michielssen, and J. Song. *Fast and Efficient Algorithms in Computational Electromagnetics*. Artech House, 2001.

[16] I. González, A. Tayebi, J. Gómez, E. García, and M. F. Cátedra. "New moment method tool combining FMLMP, CBF and MPI." *IEEE international Symposium on Antennas and Propagation and USCN/URSI National Radio Science Meeting*, 2009.

[17] F. Cátedra, L. Lozano, I. González, S. Cejudo, and C. Delgado. "New algorithm for computing antenna iterations with flat/curved structures considering any number of bounces." *30th ESA Antenna Workshop*, 2008.

[18] L. Lozano, M. J. Algar, I. González, and M. F. Cátedra. "FASANT: A versatile tool to analyze antennas and propagation in complex environments." *3rd European Conference on Antennas and Propagation*, 2009.

[19] M. J. Algar, L. Lozano, I. González, and M. F. Cátedra. "A master point strategy combined with an interpolation method to sped-up the computation of the radiation pattern of antennas on board complex structures." *ESA Workshop on Millimeter Wave Technologies and Applications and 31st ESA Antenna Workshop*, 2008.

[20] C. A. Balanis. *Advanced Engineering Electgromagnetics*. John Wiley and Sons, 1989.

[21] O. Gutiérrez, I. González, F. Saez de Adana, J. Pérez, M. F. Cátedra, and K. Vant Klooster. "Analysis of the SPORT antenna on-board the ISS using the UTD/GTD ray-tracing code FASANT." *AP 2000*, 2000.

[22] O. Gutiérrez, I. González, F. Saez de Adana, J. Pérez, M. F. Cátedra, and K. Vant Klooster. "Application of GTD/UTD method to the analysis of the SPORT antenna on board the ISS." *IEEE Antennas and Propagation Society International Symposium. Volume II*, pp. 538–541, 2000.

[23] M. Quinlan. "Galileo: A European global navigation system." *The IEE Seminar on New Developments and Opportunities in Global Navigation Satellite Systems*, 2005.

[24] F. Pérez-Fontán, V. Hovinen, M. Schönhuber, R. Prieto-Cerdeira, F. Teschl, J. Kyrolainen, and P. Valtr. "A wideband, directional model for the satellite-to-indoor propagation channel at S-band." *Int. Journal of Satellite Communications and Networking (online)*, 2009.

[25] I. González, L. Lozano, M. F. Cátedra, V. Barrena, L. Fernández, C. Carrascosa, and R. Prieto-Cerdeira. "Study of the channel propagation from satellites to indoor enviroments at L and S bands." *ESA Antenna Workshop*, 2008.

[26] E. F. Knott, J. F. Shaeffer, and M. T. Tuley. *Radar Cross Section, Second Edition*. Artech House, 1993.

[27] M. J. Algar, L. Lozano, I. González, and M. F. Cátedra. "Improvements algorithms to compute the radar cross section of electrically large complex targets considering n-bounces." *PIERS (Progress In Electromagnetics Research Symposium)*, 2010.

About the Authors

Francisco Saez de Adana received a B.S., an M.S., and a Ph.D. in telecommunications engineering from the University of Cantabria, Spain, in 1994, 1996, and 2000, respectively. Since 1998, he has been working at the University of Alcalá, first as an assistant professor and then as a professor starting in 2002. He worked as a faculty researcher at Arizona State University from March 2003 to August 2003 and as a visiting professor at the University of Technology of Sydney in 2008.

Dr. Saez de Adana has participated in more than 40 research projects with Spanish, European, American, and Japanese companies and universities related to the analysis of on-board antennas and radio propagation in mobile communication and RCS computation. He has directed three Ph.D. dissertations and has published 24 papers in peer-reviewed journals, two book chapters, and more than 60 conference contributions at international symposia. His research interests are in the areas of high-frequency methods in electromagnetic radiation and scattering, on-board antenna analysis, radio propagation in mobile communications, and ray-tracing acceleration techniques.

Oscar Gutiérrez received a B.S. and an M.S. in telecommunication engineering from the University of Cantabria, Spain, in 1993 and 1996, respectively, and a Ph.D. in telecommunications engineering from the University of Alcalá, Spain, in 2002.

From 1995 to 1998, Dr. Gutiérrez worked in the Communications Engineering Department at the University of Cantabria as a research assistant. From 1998 to 2000, he worked in the Signal Theory and Communications Department at the University of Alcalá. Since 2001, he has been an assistant professor in the Computer Science Department at the University of Alcalá. He has participated with Spanish and European companies in more than 40 research projects related to the analysis of on-board antennas, radio propagation in mobile communications, and RCS computation. He has published 13 papers in international journals, a book chapter, and more than 40 conference contributions at international symposia. His research interests are high-frequency methods in electromagnetic radiation, as well as scattering and ray-tracing acceleration techniques.

Iván González received a B.S. and an M.S. in telecommunications engineering from the University of Cantabria, Spain, in 1994 and 1997, respectively, and a Ph.D. in telecommunications engineering from the University of Alcalá, Spain, in 2004.

Dr. González worked in the Detectability Laboratory of the National Institute of Technical Aerospace (INTA) in Madrid in RCS prediction and measurements and as an assistant researcher at the University of Alcalá. Since 2004, he has worked as an assistant professor at the University of Alcalá in the Computer Science Department, teaching concepts of database systems. He has participated in several research projects with Spanish and European companies related to the analysis of on-board antennas, radio propagation in mobile communications, and RCS computation. His research interests include numerical methods applied to electromagnetic problems and rigorous and asymptotic techniques such as the MoM, GO/UTD, PO, and others. He also studies numerical methods to represent complex bodies for electromagnetic techniques and computer graphics.

Manuel F. Cátedra received an M.S. and a Ph.D. in telecommunications engineering from the Polytechnic University of Madrid (UPM) in 1977 and 1982, respectively. From 1976 to 1989, he worked at the Radiocommunication and Signal Processing Department of the UPM. He was a professor at the University of Cantabria from 1989 to 1998. He is currently a professor at the University of Alcalá in Spain.

Dr. Cátedra has worked on about 80 research projects solving problems of electromagnetic compatibility in radio and telecommunication equipment, antennas, microwave components, and radar cross section and mobile communications. He has developed and applied CAGD tools for radio-equipment systems such as those used on Navy ships, aircrafts, helicopters and satellites, with the main contractors being Spanish or European institutions such as EADS, ALCATEL, CNES, ALENIA, ESA, DASA, SAAB, INTA, BAZAN, INDRA, and the Spanish Defense Department.

He has directed 16 Ph.D. dissertations, has published close to 60 papers in well-recognized journals, two books, and close to 10 chapters in different books, and he has given short courses and more than 200 presentations at international symposia. He is also an IEEE Fellow.

Lorena Lozano received a B.S., an M.S., and a Ph.D. in telecommunications engineering from the University of Alcalá, Spain, in 2000, 2002, and 2006, respectively. Since 2005, she has worked at the University of Alcalá, first as a faculty researcher and, since 2007, as an assistant professor. She worked as a faculty researcher at Arizona State University from May 2004 to November 2004.

Her areas of interest are on-board antenna analysis, radio propagation in mobile communications, ray-tracing techniques, and high-frequency techniques. Dr. Lozano has worked on nearly 40 research projects solving problems of radar cross section computation, the analysis of on-board antennas, mobile communications, and radio propagation. She has given short courses and about 20 presentations at international symposia. She has authored four papers in peer-reviewed journals and one book chapter.

Index

Recent Titles in the Artech House Electromagnetics Series

Tapan K. Sarkar, Series Editor

Advances in Computational Electrodynamics: The Finite-Difference Time-Domain Method, Allen Taflove, editor

Analysis Methods for Electromagnetic Wave Problems, Volume 2, Eikichi Yamashita, editor

Analytical and Computational Methods in Electromagnetics, Ramesh Garg

Analytical Modeling in Applied Electromagnetics, Sergei Tretyakov

Applications of Neural Networks in Electromagnetics, Christos Christodoulou and Michael Georgiopoulos

CFDTD: Conformal Finite-Difference Time-Domain Maxwell's Equations Solver, Software and User's Guide, Wenhua Yu and Raj Mittra

The CG-FFT Method: Application of Signal Processing Techniques to Electromagnetics, Manuel F. Cátedra, et al.

Computational Electrodynamics: The Finite-Difference Time-Domain Method, Second Edition, Allen Taflove and Susan C. Hagness

Electromagnetic Waves in Chiral and Bi-Isotropic Media, I. V. Lindell, et al.

Engineering Applications of the Modulated Scatterer Technique, Jean-Charles Bolomey and Fred E. Gardiol

Fast and Efficient Algorithms in Computational Electromagnetics, Weng Cho Chew, et al., editors

Fresnel Zones in Wireless Links, Zone Plate Lenses and Antennas, Hristo D. Hristov

Grid Computing for Electromagnetics, Luciano Tarricone and Alessandra Esposito

High Frequency Electromagnetic Dosimetry,
David A. Sánchez-Hernández, editor

Iterative and Self-Adaptive Finite-Elements in Electromagnetic Modeling, Magdalena Salazar-Palma, et al.

Numerical Analysis for Electromagnetic Integral Equations,
Karl F. Warnick

Parallel Finite-Difference Time-Domain Method, Wenhua Yu, et al.

Practical Applications of Asymptotic Techniques in Electromagnetics,
Francisco Saez de Adana, et al.

Quick Finite Elements for Electromagnetic Waves, Giuseppe Pelosi, Roberto Coccioli, and Stefano Selleri

Understanding Electromagnetic Scattering Using the Moment Method: A Practical Approach, Randy Bancroft

Wavelet Applications in Engineering Electromagnetics,
Tapan K. Sarkar, Magdalena Salazar-Palma, and Michael C. Wicks

For further information on these and other Artech House titles, including previously considered out-of-print books now available through our In-Print- Forever® (IPF®) program, contact:

Artech House Publishers
685 Canton Street
Norwood, MA 02062
Phone: 781-769-9750
Fax: 781-769-6334
e-mail: artech@artechhouse.com

Artech House Books
16 Sussex Street
London SW1V 4RW UK
Phone: +44 (0)20 7596 8750
Fax: +44 (0)20 7630 0166
e-mail: artech-uk@artechhouse.com

Find us on the World Wide Web at: www.artechhouse.com